T0344442

HIGH RESOLUTION PRESSUREMETERS AND GEOTECHNICAL ENGINEERING

High-Resolution Pressuremeters and Geotechnical Engineering focuses on pressuremeters with internal transducers that make possible the derivation of strength, stiffness and *in situ* reference stress. It outlines the principles and the basics of the technology, gives guidance on good practice for obtaining and interpreting data, provides case studies and compares pressuremeter data with what is provided by other devices to enable engineers to make informed choices.

  Provides a unique, up-to-date guide to high-resolution pressuremeters with internal transducers

  Contextualises analyses and advice to enable an informed choice of testing processes

  Presents analyses previously unpublished in book form

This practical guide will suit professionals at the consultancy level, pressuremeter practitioners and site investigation companies.

HIGH RESOLUTION PRESSUREMETERS AND GEOTECHNICAL ENGINEERING

AND

GEOTECHNICAL ENGINEERING

The measurement of small things

John Hughes and Robert Whittle

CRC Press
Taylor & Francis Group
Boca Raton London New York

CRC Press is an imprint of the
Taylor & Francis Group, an **informa** business

First edition published 2023

by CRC Press
6000 Broken Sound Parkway NW, Suite 300, Boca Raton, FL 33487-2742

and by CRC Press
4 Park Square, Milton Park, Abingdon, Oxon, OX14 4RN

CRC Press is an imprint of Taylor & Francis Group, LLC

© 2023 John Hughes and Robert Whittle

Reasonable efforts have been made to publish reliable data and information, but the author and publisher cannot assume responsibility for the validity of all materials or the consequences of their use. The authors and publishers have attempted to trace the copyright holders of all material reproduced in this publication and apologize to copyright holders if permission to publish in this form has not been obtained. If any copyright material has not been acknowledged please write and let us know so we may rectify in any future reprint.

Except as permitted under U.S. Copyright Law, no part of this book may be reprinted, reproduced, transmitted, or utilized in any form by any electronic, mechanical, or other means, now known or hereafter invented, including photocopying, microfilming, and recording, or in any information storage or retrieval system, without written permission from the publishers.

For permission to photocopy or use material electronically from this work, access www.copyright.com or contact the Copyright Clearance Center, Inc. (CCC), 222 Rosewood Drive, Danvers, MA 01923, 978-750-8400. For works that are not available on CCC please contact mpkbookspermissions@tandf.co.uk

Trademark notice: Product or corporate names may be trademarks or registered trademarks and are used only for identification and explanation without intent to infringe.

Library of Congress Cataloguing-in-Publication Data
Names: Hughes, John (Engineer), author. | Whittle, Robert, author.
Title: High resolution pressuremeters and geotechnical engineering : the measurement of small things / John Hughes, Robert Whittle.
Description: First edition. | Boca Raton : CRC Press, 2023. | Includes bibliographical references and index.
Identifiers: LCCN 2022011480 (print) | LCCN 2022011481 (ebook) | ISBN 9781032060941 (hardback) | ISBN 9781032060958 (paperback) | ISBN 9781003200680 (ebook other)
Subjects: LCSH: Shear strength of soils--Testing. | Pressure transducers.
Classification: LCC TA710.5 .H84 2023 (print) | LCC TA710.5 (ebook) | DDC 624.1/5136--dc23/eng/20220711
LC record available at https://lccn.loc.gov/2022011480
LC ebook record available at https://lccn.loc.gov/2022011481

ISBN: 978-1-032-06094-1 (hbk)
ISBN: 978-1-032-06095-8 (pbk)
ISBN: 978-1-003-20068-0 (ebk)

DOI: 10.1201/9781003200680

Typeset in Sabon
by MPS Limited, Dehradun

Contents

Preface

PART I MAKING A START

In 1967, I was a junior engineer supervising the construction of fills on the rail line from Grand Prairie to the coal fields in Hinton in Northern Alberta. One of my tasks was monitoring the piezometric readings under the fill. The instruments measured the increase in pore pressure during loading. Limits were set on the pressure, relative to the embankment height, to ensure safety. The fill under construction was intended to be 32 metres high and by the end of the day was 1 metre short of this. All the pore pressure readings were well below critical.

The filling operation was continuous and proceeded throughout the night. The site engineer and I left for the town of Grand Prairie, some three hours away, leaving the contractor to finish the task. At 2.00 am the site engineer was rung up by the contractor to say the trucks could no longer see the fill. We rushed back.

The fill had failed. The top had moved down some 3 metres. The toe on the downhill side had moved out several metres. The culvert under the fill was partially open but some distance into it was cracked and split. It was just about possible to walk through the culvert from end to end. As one can imagine, this failure caused a problem that the senior railway engineers and consultants from other parts of Canada flew in to investigate. I was too junior to deal with these experts so I played about to see if I could work out for myself what had happened.

The culvert was constructed of plates stamped out to a uniform size so it was possible to determine the lateral spread. To my surprise the spread was much larger than the crack or split where the failure went through the culvert. It appeared that the entire culvert had stretched during construction. To check this, I then measured the culverts in several completed fills. They had all stretched varying amounts confirming that during construction there had been significant lateral strain. In the case of the failed embankment the movement had been enough to pinch the cables from the piezometers to the reading house and was probably the cause of the misleading readings.

We now had a simple way to monitor construction—just measure the culvert expansion. If the movement accelerated then pause work and see if it stops (slope indicators, which were then just starting to come into general use, are employed in the same way). This was useful but unsatisfactory. An engineer wants to predict the stress and strain response of his structures, not just respond to events.

At that time, we could calculate vertical settlement but not the lateral movement. That came with the introduction of Finite Element analysis. I was aware of this technique and tried to calculate what the movement might be. The difficulty was the limited data for setting up the model. The soil samples gave me an idea of the strength and stiffness but nothing about the initial lateral stress. However, a crude indication of the likely behaviour could be made of the expected movement using some of the culvert data.

At about that time, I attended a talk in Vancouver given by Prof. Roscoe of Cambridge about the work being done at the Cambridge Soil Labs on ground movement. An unconnected combination of events meant that three years later I was in that department studying under the supervision of Peter Wroth.

The Cambridge University Engineering Group had developed a technique using X ray photography that could quantify movement within a stressed soil sample pre-loaded with lead shot. Superimposing photographs taken before and after an experiment indicted the relative displacements and their distribution through the sample. We experimented with pushing sharpened tubes into prepared soil and concluded that a cylindrical probe capable of tunnelling its own way into the ground had the potential for determining the lateral stress. Such a self-boring probe would have to be a very exact fit to the cavity it was making (Fig. 0.1).

The first self boring probe we constructed had a load cell built into its surface. An external stress acting on the probe would make the cell compress. This movement was cancelled by manually adjusting the internal pressure of the device, restoring the cell to its at-rest position. Once a stable state was found the internal pressure would, in principle, be the original lateral stress.[1] The process took several hours so we investigated the possibility of using an expansion pressuremeter to get similar data much

Figure 0.1 Self-boring probe used in early experiments.

faster. Such a device would also provide a great deal more information about the stress/strain characteristics of the soil.

LOUIS MÉNARD

There was an existing device that showed the potential. The Ménard pressuremeter had been in commercial use since 1957. Louis Ménard (1933–1978) was a brilliant French engineer who had developed a pre-bored expanding pressuremeter system. For the time, it showed an innovative appreciation of the behaviour of soil. The research work was done at the University of Illinois in the United States and his company now had a presence throughout Europe and Canada (Fig. 0.2).

More than 60 years on, the concept remains a powerful one and can be used to give a mechanical picture of the response of the ground. In simple terms, his "Pressuremeter" expands a membrane against the borehole wall and measures three parameters—a limit pressure, a creep pressure and a modulus. It has been shown by many that these parameters can be related to field performance of simple structures.

The Ménard pressuremeter is expanded by injecting fluid down a hose. The pressure and the volume changes are measured at the surface. The ground moves laterally during the test rather than vertically as would be the case in a common foundation situation (Fig. 0.3).

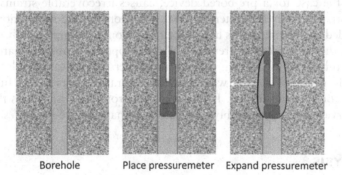

Borehole   Place pressuremeter   Expand pressuremeter

Figure 0.2 The Ménard test concept.

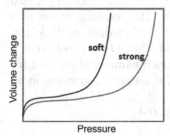

Pressure

Figure 0.3 Ménard test field curves.

What Ménard observed is that the displacement/pressure curve was distinctly non-linear and the pressure tended to a steady-state limit. Hence, working backwards, there was a relationship between this limit and the strength of the soil.

The second observation was that the initial part of the pressure expansion curve had an extended linear portion. The slope of this gave a modulus related to the stiffness of the ground.

The third observation was "creep pressure." As a consequence of the method of applying pressure and the need to write the values by hand, he observed that there was a distinct change in the amount of movement (creep) between successive steps of pressure.

Although there have been some refinements to the equipment, especially in the area of data capture, this remains the basis of the Ménard method. It is an empirical approach with some analytical background. The probe itself has no active parts. The parameters produced are specific to that method and are not fundamental soil properties.

THE SELF-BORING EXPANSION PRESSUREMETER

The attraction of a pressuremeter that can expand a cavity is the possibility of recovering the stress/strain properties of the ground. The radiographic laboratory experiments had shown that unloading the cavity, which is always the case for a pre-bored device, causes irrecoverable strains in the soil mass, especially for softer material. Additionally, the radial movement we needed to determine was below the resolution of a simple passive device such as the Ménard pressuremeter. We developed a probe built around a hollow tube, and used strain gauged transducers to measure displacement. A small version (Fig. 0.1) was built for laboratory tests and a full-scale version for the field testing. It was apparent from the first results that the field curves were considerably less disturbed than for pre-bored tests.

ANALYSIS

Limit pressure was a well-known concept in the plastic behaviour of metals subjected to extreme stress. Indeed, within the armament industry it had probably been appreciated since the mid-1800s. The potential power of a cannon was related to the barrel strength. This, rather than its thickness, was the significant parameter. Bishop, Hill and Mott (1945) had given a closed form solution for the case of a small punch hammered into a large disc of copper. Their analysis was a solution for a cavity expansion in an infinite medium. Given the elastic modulus and the yield stress of the material, the limit state could be predicted (Fig. 0.4).

Figure 0.4 Cannon power depends on barrel strength.

Gibson and Anderson (1961) used this approach to produce a pair of solutions for cavity expansion in a linear elastic/perfectly plastic soil experiencing a drained or undrained loading. The undrained solution is still widely used. Unlike the earlier solution based on the response of metals, it requires a reference stress to be specified, because the ground strength is not mobilised until the load being applied exceeds the geostatic stress. The application of the solution separates the limit pressure into its constituent parts and in particular identifies the undrained shear strength. The original solution has been refined and extended but the concepts remain sound.

The data used by Gibson and Anderson to illustrate their solution were obtained from Ménard pressuremeter tests but the analysis was independent of the equipment. The major uncertainty was knowing the initial lateral stress, the reason I had come to Cambridge. Looking at the self-bored field test data, there was a distinct alteration of slope at pressures that could conceivably be the lateral stress. Potentially, a cavity expansion solution such as the Gibson and Anderson result could now be applied to solve the field curve analytically rather than by empirical correlation and guesswork (Fig. 0.5).

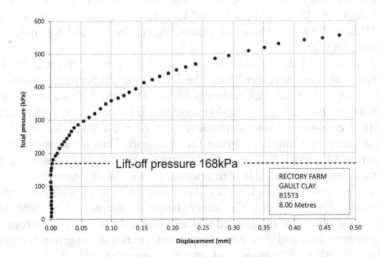

Figure 0.5 Possible lateral stress.

It was a start.

[signature: Jmo Hughes]

PART 2 WHERE ARE WE NOW?

"Everything should be made as simple as possible, but not simpler."
Roger Sessions 1950

Pressuremeters is a generic term that covers a wide variety of equipment and techniques, and it is not the purpose of this book to give a comprehensive description of this diversity. The references at the end of this preface list several works that taken together give an overview of what is available, in particular Clarke (1994). Mair and Wood (1987) is an excellent introduction to the analysis and interpretation. If the intent is to know more about Ménard technology, then it would be difficult to better Baguelin, Jezequel and Shields (1978).

All these references in their various ways try to be inclusive. It is clear now that methods and results from high-resolution pressuremeters are very different from what has been done in the past. Acting as if one is a sub-division of the other is seriously misleading to both, potentially creating unreasonable expectations of the earlier practice and limiting the possibilities of the new.

What is intended in this text is a description of the tools and analyses that best represent the developing edge of pressuremeter technology. This is a bias towards the research done by the university engineering departments of Cambridge and Oxford in the United Kingdom, Perth in western Australia and UBC in Canada. There are others, especially in the USA and, more recently, Singapore.

This is also a bias towards the pressuremeter probes invented by John Hughes. Together with Peter Wroth, he devised the self-boring pressuremeter whilst doing a doctorate at Cambridge University in the 1970s and later went on to design the high-pressure dilatometer (HPD) and cone pressuremeter (CPM). Versions of these instruments are manufactured and developed commercially by Cambridge Insitu Ltd, England. The characteristic of this generation of pressuremeter and dilatometer (for most purposes the two terms are interchangeable) is high resolution of small movement.

There are two parts to this. It is one thing to measure small movement and many devices do so. The other aspect is the ability to distinguish between movement of the ground and movement due to the finite properties of the instruments and their parts. At the extremities of the ground response (soft clay/weak rock), this second aspect becomes critical.

The Cambridge family of pressuremeters are not the only tools for which this text is relevant but most other devices that satisfy the specification are prototypes created for research purposes.

A cold look at the current state of usage of pressuremeter technology of this type would conclude that to date, high-resolution pressuremeter data are seldom used at all in the design of engineered structures. There are indications that the small strain stiffness results are beginning to be appreciated and applied. However, there is considerable caution (and misunderstanding) about the reliability and applicability of pressuremeter-determined parameters. In general, they are treated as the upper limit to any trend selected for design purposes.

All laboratory and *in situ* techniques start from the assumption that the ground is an engineering material. It is subject to the same physical rules and testability criteria as any other isotropic and homogeneous material such as a metal. The difficulty is that unlike these materials, soil gets its strength from the surrounding stress regime that may also be anisotropic in at least one plane. A better comparison to metals would be rock, where the strength comes primarily from internal bonds. A sample of a piece of metal when tested is expected to give an accurate representation of the behaviour of the whole. The ground is not that obliging—removing a sample sacrifices the stress environment and makes any subsequent test on that sample sensitive to scale factors and the hope that it retains some memory of its stress history.

Consider Fig. 0.6. This compares the amount of material influencing a typical pressuremeter test to a triaxial specimen. In terms of the volume of material that has gone plastic, the ratio between the laboratory sample and the SBP test is about 50. This is the trivial part. Of much greater importance is that the pressuremeter test works by mobilising the properties of the ground *beyond* the plastic zone. At all times the world is pushing back.

This difference of scale is why *all* pressuremeter data, even when obtained from relatively primitive equipment, should be treated with respect. It is also the reason why it can be problematic reconciling a pressuremeter-determined value for strength with what is obtained from the laboratory. There are other factors but scale is the greatest factor.

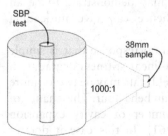

Figure 0.6 The scale effect.

The other aspect of a pressuremeter test (again, *any* pressuremeter test) is that the force applied is representative of the structural loads. Even a test in a moderately stiff clay will apply the equivalent of several tonnes to the cavity wall. It is a full-scale experiment.

The advantage of *in situ* and in particular pressuremeter testing over laboratory methods is the potential for the *in situ* test to be carried out with the ground in its natural stress condition. Whether this potential is realised in practice is one of the questions that has to be addressed in the analytical process.

It is unfortunate that data trends against depth do not include this scale difference when plotting a point. As it is, inevitably quantity of measurement over quality tends to win the battle for attention.

The aim of this book, therefore, is to get that attention. Our target is engineers who, unfamiliar with how pressuremeters produce their results, can replicate for themselves the quoted parameters using the descriptions of the analyses provided. They can then judge whether the results are relevant for the particular problem with which they are faced. The only tool required is a spreadsheet.

We are not, at this stage, concerned with design. We are assuming that our target audience will know what to do with a value for shear strength or shear modulus, once assured of the reliability of the measurement.

There are many analyses for the cavity expansion problem. Those presented are a selection sufficient for the purpose in hand, which is identifying the engineering properties of the ground to an acceptable tolerance. This means that to a large extent we are neglecting some of the more advanced analyses that solve certain technical and conceptual difficulties without greatly improving the efficacy of the primary task. This is an area that will be returned to later.

It is sometimes difficult to decide the appropriate level of complexity to apply to any problem involving the examination of data, but that is the essence of engineering. There is a caution. This part of the preface starts with a quotation usually (and incorrectly) attributed to Einstein.[2] It is indeed possible to be too simple. The pressuremeter test does not directly measure *any* fundamental property of the ground. What it provides is a loading curve that can be solved using analytical techniques. These are reliant on assumptions that later knowledge may demonstrate to be an insufficient representation of reality, and hence can give misleading predictions of certain parameters.

An example is the assumption of a stress/strain response that is simple elastic/perfectly plastic. All soils exhibit a non-linear stiffness degradation characteristic before yielding. Stiffness at the yield state may be two or more orders of magnitude less than the very small strain behaviour. This change of perspective has major implications for a number of cavity expansion solutions, such as Gibson and Anderson (1961). In this case it doesn't make the solution unusable for determining strength, but it does make conclusions about the yield strain invalid. There is some skill in separating what is still sensible from what is now questionable.

THE NEED TO GO SMALL

A pressuremeter is one of a limited number of tools with the potential for deriving representative parameters for the engineering properties of soil and rock. It is unique in being able to provide strength and stiffness in a single test episode. These parameters are identified by solving the boundary value problem.

It is essential to the application of any solution that the stress and strain at which the material yields are correctly quantified. Fortunately, it is not necessary to determine the elastic properties from the initial part of the test. They can be discovered at any point in the expansion process by reversing the direction of loading and stress cycling the material a small amount (see Fig. 0.7 for an example). This simple but powerful technique gives access to an astonishing amount of information concerning the sub-yield characteristics of the ground—and it is repeatable. All that is necessary is to be able to see very small steps of change. Implicit in this is the need for data to be quiet and plentiful. The test needs to be defined by hundreds if not thousands of such measurements.

Once the yield properties of the ground are known, then the solutions can be applied in a way that not only determines strength and stiffness but quantifies the damage done to the material during the insertion process. The complete interpretation is in principle a coherent narrative in which all parts of a test are accounted for.

PASSIVE PRESSUREMETERS

If it cannot be seen, then it cannot be used. Pressuremeter systems without transducers in the probe itself, such as a Ménard device, are unable to determine strains at the level required and their data are not amenable to treatment in this forensic way. It was logical and probably commercially astute of Ménard to bypass the analytical difficulty by developing a practice based on correlations.

However, reverting to empiricism reduces the significance of the individual test. Essentially a Ménard result is a superior version of a blow-count obtained from a Standard Penetration Test (SPT). Like the SPT, multiple tests at intervals not much greater than the length of the probe are required to minimise the uncertainty and a rigid test and interpretation procedure becomes mandatory.

Because of this method specificity, there is no incentive to innovate the probe. It remains much as it did at its inception in 1957, a passive measuring device that makes comparatively coarse determinations of volume change and pressure.

The Ménard approach will not be considered any further here, although it is as well to be aware that the overwhelming majority of pressuremeter tests carried out still use equipment and methods based on Ménard[3] practice.

It is also appropriate to point out that two of the best known analyses for the cavity expansion problem, Gibson and Anderson (1961) and Palmer (1972) both use Ménard data to demonstrate their solutions.

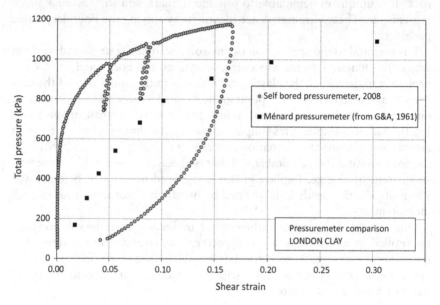

Figure 0.7 Low and high resolution data compared.

Fig. 0.7 shows two field curves plotted on the same scales. One is a Ménard test, used as an example by Gibson and Anderson in their paper. About 50 years later, a Cambridge Self-Boring Pressuremeter was deployed in the same material and at the same level.

For the purposes of deriving strength and limit pressure both data sets give similar answers. These are parameters identified from the plastic response. The obvious difference is the total absence of any elastic information in the earlier test.

It's the difference between a magnifying glass and a microscope.

PRESENTATION

Most of the examples in this book are the output of an analysis program, WINSITU, that has been in use, in one form or another, for over 30 years.

Each chapter is arranged to be as far as possible a stand-alone document, with its own notation and references. This portability leads to a certain amount of repetition. Like previous authors, we have tended to break down the subject by separate parameters, such as strength, stiffness and permeability. This is necessary but unfortunate as it masks the interconnectedness of these different ways of looking at the same data. The sequence of chapters replicates the usual order of the analysis process.

NOTES

1 Later versions of this device use an automatic feedback system to find the null position.
2 What he actually said, possibly, was "It can scarcely be denied that the supreme goal of all theory is to make the irreducible basic elements as simple and as few as possible without having to surrender the adequate representation of a single datum of experience."(lecture, 1933). Not quite as pithy.
3 Louis Ménard met with and helped John Hughes in the early stages of his research.

REFERENCES

Baguelin, F., Jezequel, J.F. & Shields, D.H. (1978) *The Pressuremeter and Foundation Engineering*. Transtech Publications, Clausthal, Germany ISBN 0-87849-019-1.

Bishop, R.F., Hill, R. & Mott, N.F. (1945) The theory of indentation and hardness tests. *Proceedings of the Physical Society* **57** (3), pp. 147–159.

Clarke, B.G. (1994) *Pressuremeters in Geotechnical Design*. Publ. Taylor & Francis Ltd, ISBN: 9780751400410.

Gibson, R.E. & Anderson, W.F. (1961) In situ measurement of soil properties with the pressuremeter. *Civil Engineering and Public Works Review* **56** (658), pp. 615–618.

Hughes, J.M.O. (1973) An instrument for in situ measurement in soft clays, PhD Thesis, University of Cambridge.

Mair, R.J. & Wood, D.M. (1987) Pressuremeter testing. Methods and interpretation. *Construction Industry Research and Information Association Project 335*. Publ. Butterworths, London. ISBN 0-408-02434-8.

Palmer, A.C. (1972) Undrained plane-strain expansion of a cylindrical cavity in clay: a simple interpretation of the pressuremeter test. *Géotechnique* **22** (3), pp. 451–457.

Author bios

Frustrated by the lack of tools for measuring the lateral geostatic stress, **John Hughes** went to Cambridge University from Canada in 1970. Together with Peter Wroth he developed the Self Boring Pressuremeter and solutions for interpreting cavity expansion tests in clays and sands. His 1977 paper on sands is a seminal work. For more than 50 years he has combined academia with practical engineering. He has performed field testing in almost every continent and some of the remotest locations, using *in situ* testing equipment. It is a fundamental aspect of his approach that ground movement be measured in the smallest possible increments.

Robert Whittle began working for Cambridge Insitu (CI) in 1978 as an electronics engineer and gradually moved towards data interpretation and writing software for the purpose. He has spent over 40 years carrying out field testing world-wide and competes with John on the quantity of tests performed and countries where he has worked. He is the author and co-author of many journal and conference papers concerning pressuremeters and making sense of the data they produce. Since late 2018 he has been co-owner of CI.

Chapter 1

Introduction

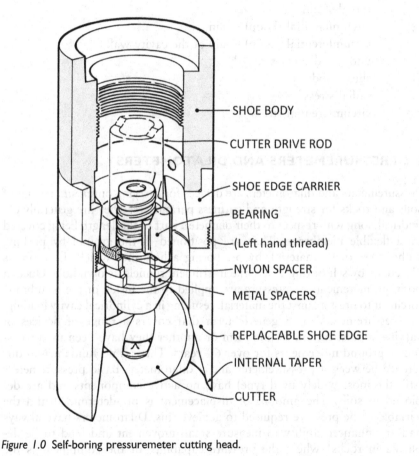

Figure 1.0 Self-boring pressuremeter cutting head.

"All models are wrong, but some are useful."

George Box, 1978

1.1 NOTATION

ρ	a small radial displacement
r	any radius
r_c, r_y	radius of cavity, radius of elastic boundary
r_o	initial radius of cavity
p_c	total pressure applied to cavity wall
p_o	cavity reference pressure
p_y	yield stress (at the elastic/plastic transition)
P_f	total yield stress
ε_r	radial strain
ε_θ	circumferential (hoop) strain
ε_c	circumferential (hoop) strain at the cavity wall
c_u	undrained shear strength
G	shear modulus
σ_r	radial stress
σ_c	circumferential stress

1.2 PRESSUREMETERS AND DILATOMETERS

Pressuremeters and dilatometers are devices for carrying out *in situ* testing of soils and rocks for strength and stiffness parameters. They are generally cylindrical, long with respect to their diameter, part of this length being covered by a flexible membrane. They are positioned in the ground by pushing without removing material, by pre-boring a hole into which the probe is placed, or by self-boring where the instrument tunnels its own hole. Once in position, increments of pressure are applied to the inside of the membrane forcing it to press against the material, resulting in a cylindrical cavity loading.

"Pressuremeters" is a generic term that covers a range of devices of varying sophistication. In one form or another they have been used to estimate ground movements for over 60 years. There is no fundamental difference between a pressuremeter and a dilatometer. Basic pressuremeters (still the most widely used type) have no active components and are deployed in soils. The amount of displacement is pre-determined and the variable is the pressure required to achieve this. Dilatometers have always had an enhanced ability to measure small movement and tend to be deployed in rocks[1] where the pressure capability of the equipment is the limiting factor. These distinctions are no longer relevant.

A pressuremeter test consists of a series of readings of pressure and the consequent displacement of the cavity wall (Fig. 1.1). The loading curve so obtained may be analysed using solutions for cavity expansion and contraction in an infinite medium. A test is usually carried out in a vertical hole so the

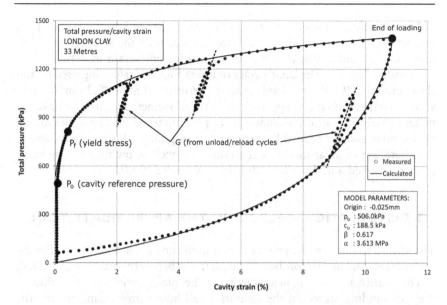

Figure 1.1 Undrained self-bored pressuremeter test in a stiff over-consolidated silty clay.

derived parameters are those appropriate to the horizontal plane. The fundamental aspects of these solutions have been known for well over 150 years.

Cavity expansion and contraction theory is used in ballistics and biomechanics as well as in the geotechnical field. It can calculate the penetration of a projectile into concrete or the rate of growth of a tumour in the brain. Other examples of cavity expansion and contraction processes in the geotechnical world are the cone penetration test (CPT) and the standard penetration test (SPT). Certain problems such as the movements affecting the construction of shafts, tunnels and the short and long-term behaviour of piles are explicitly understood in relation to cavity expansion theory. Some pressuremeter field data, combined with the particular means of inserting the probe (pushed or pre-bored) are a direct analogue for these configurations. However, for the most part, the pressuremeter test is arranged to produce a loading curve that can be deconstructed to find the underlying shear stress/shear strain response of the ground.

A sheet of paper has a thickness of about 0.1 millimetres. The instruments that we are concerned with can resolve changes of displacement at least 100 times smaller than this, about the wavelength of ultraviolet light. They are electro-mechanical devices with multiple transducers carrying out local measurement of pressure and displacement to a high level of discrimination. It is routine to measure shear strains of 0.005%; to put this into context, shear failure occurs at about 1% in clays and is even less for a sand. The maximum shear strain that the instruments can apply varies between 25% (for the self-boring probe) and 100% for more intrusive devices.

The remarkable aspect of cavity expansion theory is this—if the stress and strain required to induce yield are known, then the stress required to move the ground 100% is also known, and for any strain in between.[2]

Consider Fig. 1.1. The small cycles of unloading and reloading used to find shear modulus (G) have a displacement alteration of about 0.2 mm (or two sheets of paper) and this is approximately the movement required to make the material yield, the strain ordinate of the point marked p_f. The cycles are all very similar and are defined by a smooth set of points. This regularity allows the alteration of shear modulus at strain levels well below the yielding value to be quantified and described (Bolton & Whittle, 1999).

1.3 DISTURBANCE—SELECTING THE APPROPRIATE TOOL

The ideal is to position the pressuremeter in the ground in such a way that the state of stress around the probe remains as it was before the instrument arrived.

The reality is that no instrument can be placed without alteration of the surrounding stress. In the case of a self-bored pressuremeter test, the disturbance is less than that required to make the material adjacent to the probe become fully plastic (yield). This degree of disturbance is straightforward to quantify and allow for as part of the analysis procedure.

This is an unusual situation in the context of soil investigation. Most testing, including pressuremeter testing, involves more disruptive ways of getting an instrument into the ground. In soils, this means the material will be failed in extension or contraction, and sometimes combinations of both. Once the mobilised shear stress exceeds the yield condition the material is damaged beyond recovery. The loading path followed will no longer be that of the ground in its natural stress state. The extent of that difference has to be assessed and used to inform the interpretation.

It is neither sensible nor realistic to attempt to self-bore all materials. Of the alternatives to self-boring, the least damaging is placing the probe in a pre-bored or cored pocket. For some materials, especially rock, this is the only viable option. There are some compensations because core can be obtained at the same horizon as the pressuremeter test. This may be more of a theoretical advantage, as pressuremeter testing tends to be applied at precisely those places where core recovery is uncertain or of questionable quality.

In weaker material, the complete unloading of the cavity following removal of the boring tool causes reverse plastic failure. The material local to the cavity wall will be in a different condition to the undisturbed material and will remain so. If the area of material under load eventually exceeds the area of the disturbed soil by a sufficiently large amount, then it is possible to register loading data that are predominantly a function of the undisturbed ground. A large expansion capability is required if this is to be achieved.

However, if the material has been pushed, then no amount of expansion will ever follow the true stress/strain path. Any insertion process into virgin ground that starts with a point causes the soil to experience infinite strain. This is the case for all devices that push or hammer their way into the ground, such as a CPT or SPT. The pressuremeter inserted in this way has the same difficulty. This limitation is overcome by expanding the cavity even further until a limiting stress condition is restored but for an indeterminate strain.

At this point, the direction of loading is reversed. Coordinates of pressure and displacement seen at the cavity wall will be a function of remote material beyond the elastic/plastic transition. This material is being unloading elastically and eventually plastically as the contraction continues. Although the loading strains are undefined, the unloading strains are easily calculated and hence amenable to analysis.

1.4 UNLOAD/RELOAD CYCLES

This stress reversal is the basis of the unload/reload cycles that are a feature of all the field curve examples (i.e. Fig. 1.1). Because the test is conducted in an infinite soil mass a cycle taken at any expansion will give a near identical response. It is the repeatability of this technique that makes it so powerful, and in the context of ground testing, exceptional.

Fig. 1.2 represents a state of stress and strain that is never achieved in practice. A perfectly installed pressuremeter has expanded a cavity to a radius r_c by applying a pressure p_c. The material is about to yield, hence r_c is also the radius to the elastic/plastic boundary r_y. As yet, no plasticity has been initiated. The cavity pressure p_c equals the yield stress p_y. If p_c is now reduced, the elastic ground extending to infinity will respond by forcing the cavity to unload elastically. The cavity can be reloaded and r_y will continue to equal r_c provided that the internal pressure remains below the yield stress. As this is an elastic response, the displacements and pressure changes are a direct function of the shear modulus of the ground.

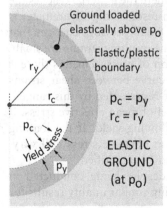

Ground loaded elastically above p_0

Elastic/plastic boundary

r_y

r_c

p_c

Yield stress

p_y

$p_c = p_y$
$r_c = r_y$

ELASTIC
GROUND
(at p_0)

Figure 1.2 Material at the point of yield.

The yield stress is the maximum that the ground can develop. If the cavity pressure p_c exceeds this, then the ground fails and a zone of plastic material develops around the pressuremeter. The yield stress now acts on a larger circumference and balances the greater force acting on a smaller circumference at the cavity wall. At all times $r_c p_c = r_y p_y$ (Fig. 1.3).

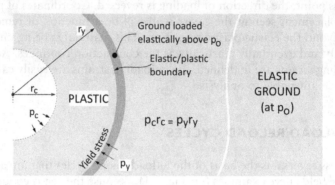

Figure 1.3 The expanded cavity.

If p_c is now reduced, the same process as illustrated in Fig. 1.2 takes place at the new yield boundary r_y. The material lying beyond r_y that was about to yield is unloaded elastically and r_y reduces. This is experienced at the cavity wall as a reduction in r_c.

It is of little consequence what has been done to the cavity itself. The magnitude of the plastic zone is unimportant and need not be known. Provided that r_y is the largest it has been, the changes to p_c and r_c are identical to those that would have occurred in the early stages of a test if a perfect installation such as that in Fig. 1.2 were achievable.

It is important that this concept be fully understood because it is central to much of what follows. "Elasticity" has been applied rather loosely to this description. It is not true elasticity (obeying Hooke's law) but a process of stiffness reduction that terminates at the point where the material reaches the maximum shear stress it can support. The high-resolution pressuremeter is easily capable of capturing this response using small pressure changes that combine to make the unload/reload envelope. The resulting strains are small, but given the resolution of the current generation of pressuremeters there is no difficulty providing a stiffness degradation curve between shear strains 0.01% and 1%. Almost all design problems are represented by this range.

Stiffness is also dependent on the current level of mean effective stress. If the test is undrained, then this is constant following yield. If the loading occurs under drained conditions, then the mean effective stress will vary. Several cycles will be required if this effect is to be quantified.

This technique means that a pushed test, despite being conducted in heavily disturbed material, is capable of providing similar quality results for stiffness and strength as a self-bored test.

1.5 WHY SELF-BORE?

If the results are similar, then why self-bore at all? There are two primary reasons for using self-boring instead of more invasive techniques. The first is that to a large extent the initial stress state is preserved and the possibility of directly determining the horizontal geostatic stress σ_{ho} becomes available. There is less dependence on the appropriateness of the analytical solution.

The second and probably stronger motive for self-boring is the impossibility of obtaining data for a troublesome zone of material in any other way. Some materials cannot be cored without losing their structure so completely as to make any subsequent sampling worthless for strength evaluation. Similarly, the range of soils that can be pushed is limited. The advantage of self-boring is that although originally developed to test soft clay and loose sand, it can be arranged to penetrate ground up to the strength of weak rock.

At the other extreme, self-boring can make a test in material such as mine tailings. These generally have a c'-ϕ' characteristic. It is essential that the friction angle ϕ' at first yield be measured because the difference between this and the critical state friction angle ϕ'_{cv} decides the liquefaction potential of the material. It is extremely difficult to obtain usable core from such deposits and a pushed pressuremeter relies on the contraction data for soil parameters; ϕ' makes a reduced contribution to the material unloading which for the most part is controlled by ϕ'_{cv}.

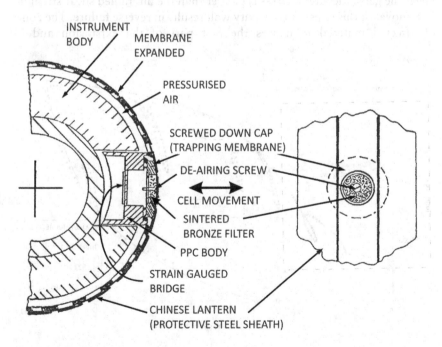

Figure 1.4 Pore pressure transducer.

The elastic membrane of the SBP has the thinnest wall thickness as a proportion of the overall diameter of all the pressuremeters. This allows it to have pore pressure transducers fitted (Fig. 1.4). The membrane is trapped between the transducer and its filter and the devices respond to the development of excess pore water pressure (or absence of it). This facility enhances the understanding of the material response to loading and makes it possible to carry out consolidation testing using the fixed strain method.

1.6 THE PRE-BORED TEST IN SOIL

The majority of pressuremeter tests make use of pre-bored pockets. The hole is drilled to the size of the core barrel or drill bit, selected to be up to 5% greater than the diameter of the probe. The amount of over-drilling will depend on the type of material, the drilling fluid and the mechanics of the drilling. A pocket is formed, long enough to accommodate comfortably all of the expanding part of the pressuremeter. The drilling tools are withdrawn and the pressuremeter is placed in the pocket.

In between these two events, there is a period when the cavity wall is unsupported and will move inwards by an amount that depends on the material properties. Fig. 1.5 is an example of a pre-bored test in a very stiff clay. This shows data plotted and analysed in a similar way to Fig. 1.1. Here, the geostatic lateral stress is greater than the undrained shear strength so removal of this stress at the cavity wall results in reverse failure. The zone of affected material influences the first part of the expansion and is

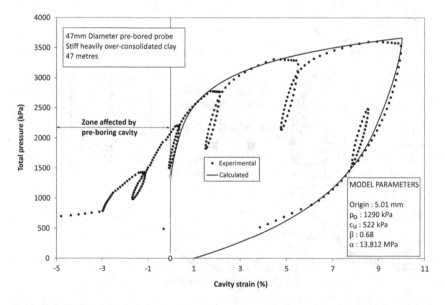

Figure 1.5 Undrained pre-bored pressuremeter test in very stiff over-consolidated clay.

irrecoverable disturbance. Nevertheless, as the example shows, if the cavity is expanded sufficiently, then the contribution of the stress relieved zone reduces and the model allows a plausible match to the latter part of the field curve to be obtained. The initial data points are ignored—it is possible to use the solution to calculate the position of these data also and in that sense the parameter set remains unique. Notice that the first cycle is very different from the second and subsequent cycles. This indicates that at the time the first cycle was initiated, the stress at the cavity wall was below that required to replace the unloading damage. By the time the second cycle was taken, this was no longer a problem.

1.7 THE PUSHED TEST

Many uniform soils will permit a tool to be pushed or hammered into place. The disturbance caused by this is total and is never recoverable regardless of how far a cavity expansion is extended. There are several variants of the pushing method, but typically the foot of the probe is fitted with a point or cone and the push into virgin ground causes the material to be taken to a limit condition. As material flows around the shoulder of the cone, there will be a partial unloading from this limit state. Consequently, the material around the pressuremeter and for some distance from it will be in a complex state of stress prior to the expansion of the pressuremeter commencing.

The purpose of the expansion phase of the test is to erase the history of the insertion process by applying a level of stress at the cavity wall that can only be contained by increasing the area of material that has gone plastic. This extends the boundary where the material is on the point of yielding to a greater radius than existed following cavity formation. All soil elements within the plastic zone are put into a uniform state and, in particular, all displacements become radial. If the direction of loading is now reversed the ground behaves as described by Fig. 1.3. If the unloading continues beyond the material elastic range it will yield in the reverse sense and a plastic response will be obtained.

Fig. 1.6 is an example from a pushed test in clay. The limit state is largely recovered after a relatively modest expansion of 5%. The direction of loading is then reversed to take the first unload/reload cycle. This procedure is repeated for two more cycles before eventually unloading completely. This contraction phase also includes a reload/unload cycle.

Fig. 1.7 shows the analysis of this type of test. The loading data contribute nothing apart from the measured limit pressure but undrained shear strength can be identified from the contraction response. If the contraction starts from the limit pressure, then it is also possible to discover the cavity reference pressure, p_o.

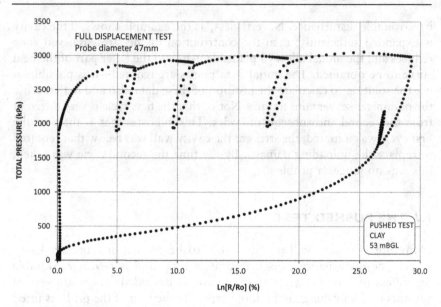

Figure 1.6 FDPM field curve.

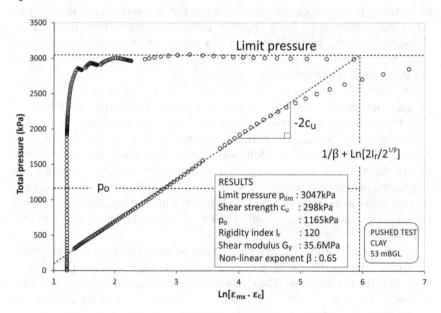

Figure 1.7 FDPM test analysed.

The published method (Houlsby & Withers, 1988) is based on a simple elastic/perfectly plastic soil model and consequently over-estimates p_o. A modification suggested by Bolton and Whittle (1999) assumes the soil response is non-linear elastic/perfectly plastic. Incorporating non-linearity in the form of a single value, β, allows credible estimates of p_o to be made. The measured data are unchanged but a small increase in the realism of the soil model greatly improves the interpretation.

Unlike the self-bored or potentially the pre-bored test, the results obtained from the analysis in Fig. 1.7 do not allow coordinates of stress and displacement from the loading phase of the measured field curve to be predicted.

The probe used for the pushed test in Fig. 1.6 is the same device as that used to obtain the pre-bored test in Fig. 1.5. The difference in the profile of the field curve is solely the consequence of how the hole was formed. If, additionally, the instrument is fitted with a live CPT it is known as a cone pressuremeter (CPM). Such an instrument combines the soil profiling potential of the CPT with the ability to measure strength and stiffness directly. If it is fitted with a dummy point, then it is more generally referred to as a full displacement pressuremeter (FDPM). The probe itself does not change, it remains the same device for all these configurations.

1.8 TESTING ROCK

For geologists, all parts of the Earth's crust are rock—soil is merely the end state of sequences of erosion acting on rock particles.

The distinction between rock and soil is an engineering construct where "rock" refers to material whose natural condition is a solid mass of bonded minerals. The strength of the bonds determines the response that is seen when rock is put under load. Because of the intact structure, water flow is primarily through discontinuities and the effective stress concepts of soil mechanics are not applicable.

For a cavity expansion in such material the primary distinction is between tests that show plastic development i.e. exceeding the strength of the bond, or tests where the material remains intact at the limits of the load that the pressuremeter can apply. Both types of test are shown in Fig. 1.8.

For the pressuremeter rock test there is both additional complexity and simplification. The possibility of tensile failure must be included in the interpretation but issues such as non-linearity in the elastic response are of limited concern. Fig. 1.8 shows high-pressure pre-bored tests in mudstone, taken from the same borehole. Nearer the surface, the rock is weathered and relatively weak, and the test shows substantial plastic deformation. At deeper depths, all that can be seen is elastic movement. However, as the plot on the right shows, if the data are examined more closely there is a plethora of detail, including four cycles of unloading and reloading. This is all within a total cavity wall movement of 0.30 mm.

Data are recorded every few seconds throughout the test and each pressure increase of 1 MPa is sustained for one minute to allow time-dependent effects to become evident. A creep displacement is the difference between the readings of the cavity radius at the start and end of a pressure hold. The form of the creep response can be helpful in identifying the onset of fracturing.

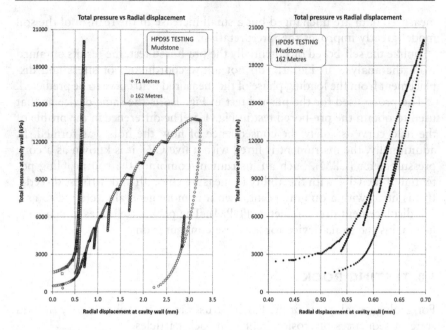

Figure 1.8 High-pressure dilatometer tests—with and without shear failure.

The problematic part of interpreting tests in rock is identifying the initial stress state. The process is essentially one of inspection of the loading curve for tiny detail that indicate, for example, instances of fracturing (Fig. 1.9). Back-calculating from the radial stress at which fracturing occurs can set an upper limit to the *in situ* lateral stress.

Despite the small overall movement, there are clear indications of fracturing between 12 and 14 MPa. It is also possible to see that of the three reloading cycles the stiffest result is obtained from the first cycle, taken before fracturing affected the response.

In addition to having the capability to measure small movement, tests in rock make considerable demands on the ability of the equipment to quantify its own finite stiffness. It is frequently the case that movements due to system compliance are of the same if not greater magnitude than the movement of the rock itself.

As a mechanism, the probe should be as stiff as possible, but it will still be only about as stiff as a weak rock. The type of instrument used to make the tests in Fig. 1.9 has a stiffness equivalent to a shear modulus of 5 GPa. This is determined with a calibration procedure that not only quantifies the probe contribution but also the repeatability of the calibration. The calibration repeats with a maximum 10% variation. Hence, the instrument can easily determine the shear modulus of rock of more than 20 GPa with only minor uncertainty, well within 20%.

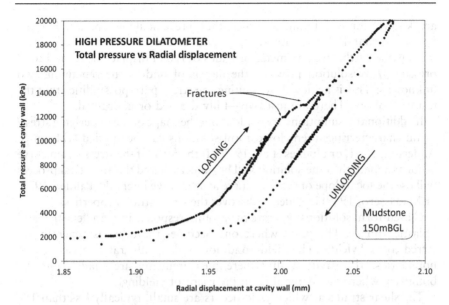

Figure 1.9 Elastic material showing fracturing.

If the material yields and irrecoverable deformations develop, then it is possible to analyse a test in much the same way as a test in particulate material. The solution may be more complex because of the need to account for tensile as well as shear strength. If the yield stress is not reached, then the analysis is limited to providing estimates of shear stiffness, setting a minimum estimate of the strength, and possibly some insight into the lateral components of the geostatic stress if time-dependent ("creep") data are collected.

1.9 CAVITY EXPANSION AND THE ANALYTICAL APPROACH

The purpose of the interpretation is to reduce the field data to parameters that describe the engineering properties of the ground. The cavity expansion test is a shearing process so the parameters in question are shear modulus, shear strength or friction angle as appropriate, and the *in situ* lateral stress.

Simplifying assumptions are made about the nature of the test and the ground. For tests in soil, it is assumed that the material is fully saturated, homogenous, isotropic and behaving as a continuum that fails in shear only. Additionally, the analytical problem is considerably simplified by assuming that the cavity is sufficiently long with respect to its diameter for axial movements to be negligible.

Consequently any solution has only to consider the radial and circumferential strains ε_r and ε_θ, and the radial and circumferential stresses σ_r and σ_θ. Given any three, the fourth can be calculated. If the four parameters are known for one boundary, then they are calculable for all. For the boundary that is the cavity wall, ε_θ and σ_r are obtained directly from measurements the pressuremeter provides of radial displacement[3] and total pressure. Any solution provides the means of finding the remaining two unknowns. This is achieved by limiting the stress path possibilities that the test can follow. There are just two—fully drained or undrained.

If additional assumptions are made about the shape of the underlying stress/strain characteristics then closed form solutions can be applied (Gibson & Anderson, 1961, or Hughes et al., 1977). If the form of the stress/strain curve is not assumed, then the solution will be numerical and the analytical process will use the local slope of the field curve at successive intervals (Palmer, 1972, or Manassero, 1989) in order to discover the stress/strain properties.

Closed-form solutions separate the ground response into an "elastic" and "plastic" phase. The point where one becomes the other is generally referred to as "yield." The yield condition is then integrated between two boundaries: the cavity wall (where measurements are made) and the boundary where the material is on the point of yielding.

The shear strains at which yield occurs are small, typically less than 1% and for the majority of tests will be masked by insertion effects. An error in this initial calculation affects all subsequent calculations, although to a diminishing extent as the test develops.

Every analytical solution is a simplified description of a complex material and loading situation. For the type of engineering situation where high-resolution pressuremeter testing is specified, the analytical approach may be only the first step in modelling the ground response and the interaction between the ground and structures. In that sense, predicting the pressuremeter field curve will be a necessary part of calibrating the geotechnical numerical model.

The appealing aspect of closed-form solutions is that they are reversible. Given a set of parameters, the solutions predict the field curve. This is the basis of the calculated trends in Figs. 1.1 and 1.5.

There are a class of complex cavity expansion and contraction solutions (for example, Yu & Houlsby, 1991, 1995). These pay particular attention to the size of the deformed region and make appropriate calculations of the large strains involved. They can be harder to implement because each solution requires solving a convergent infinite series to an acceptable accuracy. In practice, this is not too onerous because convergence is achieved after a small number of iterations.

Certain solutions make an important difference to the understanding of an analytical problem and cannot be ignored. Carter et al. (1986) allows for cohesion as well as friction and is therefore a significant advance on the purely frictional solution of Hughes et al. (1977). This in turn introduced dilatancy and was a major step forward from the earlier frictional analysis of Gibson and Anderson (1961).

Yu and Houlsby (1991) is elegant, but its improvements are not like these (the solution reduces to the Carter et al. solution if the strain definition is simplified). Unless a solution greatly improves on the quality of the predictions made for engineering parameters, then ease of use has to be an important consideration.

1.10 PRINCIPAL STRESSES AND THE FIELD CURVE

The connection between principal stresses and the field curve is sketched in Fig. 1.10 for a pressuremeter test in soil. The example is showing simple elastic behaviour before failing in shear. The stress path followed by an element of soil adjacent to the cavity is given in the left-hand sketch and the corresponding pressure/strain curve is shown alongside.

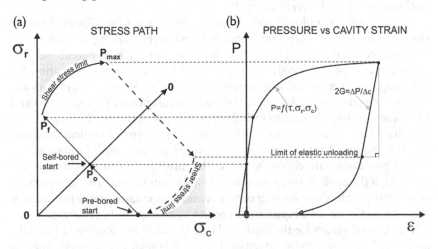

Figure 1.10 Stress diagram for a test showing shear failure.

The radial stress, ideally at the *in situ* horizontal stress for a perfect installation, increases at the same rate as the circumferential stress decreases, regardless of whether the material is deforming under plane strain or plane stress conditions. The line 0–0 represents stress equality, so the point P_0 is the *in situ* lateral stress σ_{ho}.

Once the radial stress increases above P_0, then the shear stress in the soil at the cavity wall will increase. If P_0 is low, then it is possible that the circumferential stress would go into tension. Tensile failure, as is commonly seen in rock, implies comparatively low values for the initial stress state compared to the shear strength.

The pressure necessary to initiate shear failure is denoted pf. After this pressure, the current slope of the field curve reduces steadily. The form of this part of the pressure/strain curve is a function of the shear strength properties of the material.

Following the initial shear failure, radial stress and circumferential stress increase together. If the shear stress limit is constant, and is not influenced by pressure, and if the material deforms at constant volume, then the failure shear strength can be determined by the analytical solution developed by Gibson and Anderson (1961). If the shear stress limit increases as the loading develops then the solution of Hughes et al. (1977) could be applied to discover the internal angle of friction and dilation.

Prior to reaching the shear stress limit, the pressuremeter response is elastic, both in loading and unloading. Assuming the soil deforms at a constant modulus and the installation is perfect then the slope of the initial loading path gives the shear modulus of the material, using the classic procedure of Bishop et al. (1945). The diagram also indicates that reversing the direction of loading gives a response that is initially elastic and this is the justification for unload/reload cycles.

As indicated, the unloading of the pressuremeter is a function of the same strength and stiffness properties that controlled the loading response.

From the right-hand side of the stress diagram, it is apparent that the pressuremeter provides only a limited set of the necessary information for resolving the stresses and strains around the probe. Specifically, it gives the changes in radius of the borehole wall (a particular case of hoop strain) and the corresponding changes in radial stress at the borehole wall. There are no data for hoop stress or radial strain or movements in the vertical direction and the analytical solutions must account for the missing information.

Test procedures are chosen to allow the missing data to be inferred—for example, if the application of pressure is fast enough for the material to behave as incompressible, then shearing occurs at constant volume and hence changes of radial strain must be equal and opposite to changes in hoop strain.

The stress diagram for the entirely elastic ("rock") test is given in Fig. 1.11 and includes the possibility of tensile failure. If cracking is apparent, then as

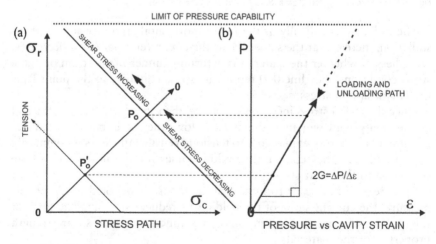

Figure 1.11 The pressuremeter test in rock.

the left-hand figure implies, assumptions about the magnitude of the cavity reference pressure can be deduced from the radial stress at which it occurs (i.e. $p_0 < \sigma_r/2$).

1.11 MEASUREMENT CONSIDERATIONS—USING THE MEAN

Precision is the degree of variation in repeated measurements of the same physical quantity in unchanged conditions. For the devices being considered, an individual displacement sensor has a precision of about 0.6 µm.

For a n-axis measurement system, combining the output of all displacement sensors and assuming a random distribution of error in the current reading means the precision will be n times better than for a single sensor. For a three-axis SBP, this makes the precision 0.2 µm.

A similar argument applies to resolution, although taking advantage of this would require reading the output to one more decimal place.

All things being equal, averaging the output from the displacement sensors gives the highest precision data. However, the most pressing argument for using the mean is that this is the measure that cancels out movements of the reference (Whittle, 1993). The probe body is the origin for all measurements, so if its axis moves laterally then from the perspective of an individual displacement transducer the cavity appears to move. Only averaged data are immune and are the preferred choice in almost all circumstances.

One exception is a self-bored test when, after boring, the probe is locked in position by the external stress and its own internal pressure has yet to exceed that stress. In such circumstances, the response of the separate displacement sensors may give indications of the initial stress state in the surrounding soil, allowing an assessment of the degree of insertion disturbance (Whittle et al., 1995).

There is a separate argument that all available analyses assume isotropic properties in the surrounding soil, and only the average pressure/strain curve represents this condition.

1.12 STRAIN DEFINITIONS

The time honoured way of describing the expansion of the cavity, when logging the data, is to refer any movement to the initial radius of the probe, expressed as a percent. Unfortunately this is frequently labelled as "radial strain", which it is for the instrument but not for the cavity, causing some confusion. For the most part, therefore, the examples throughout this text use displacement in mm for field data. Analysed data use one of the forms of strain given below to describe the expansion.

1.12.1 Simple strain

For a pressuremeter measuring the radius of an expanding cavity, the conversion from displacement to simple strain is:

$$\varepsilon_c = [r_c - r_0]/r_0 \qquad\qquad [1.1]$$

where
  r_c is the current radius of the cavity
  r_0 is the original radius of the cavity in the *in situ* state.
  ε is normally written ε_c to denote cavity strain. This is a particular case of circumferential or hoop strain. ε_c is often expressed as a percentage and by convention increases in the positive direction as the cavity enlarges. The physical reality is that hoop strain reduces as the cavity radius increases.

For a self bored cavity r_0 can be approximated by the at-rest radius of the instrument. This is unlikely to be the case for a stress relieved pocket. In general the approach then is to identify when the applied pressure has reached the in situ lateral stress, and interpolate from this the corresponding radius, which becomes r_0. As the analysis process goes through stages of iteration, r_0 is likely to be re-defined again.

1.12.2 Current cavity strain

Current cavity strain is given by:

$$\varepsilon_c = [r_c - r_0]/r_c \qquad\qquad [1.2]$$

using the same terminology as above. This takes some account of alterations to the length of the displacement reference.

1.12.3 True or natural strain

If strains are small enough then the variation in the length of the displacement reference is insignificant. This is not the case for a pushed pressuremeter test which will result in the material experiencing very large strains. It is necessary then to use true or natural strain to describe the cavity deformation. This is the sum of the incremental increase in radius divided by the current radius:

$$\varepsilon = \ln [r_c/r_0] = \ln [1 + \varepsilon_c] \qquad\qquad [1.3]$$

When carrying out unload/reload cycles, the calculation for shear modulus G is approximately

$$2G = \left[\frac{\Delta p_c}{\Delta \varepsilon_c} \right] \qquad [1.4]$$

Mair and Wood (1987) point out that this approximation is only justified for strains close to the origin and [1.4] should take account of the changing cavity radius:

$$2G = \left[\frac{r_c}{r_o} \right] \left[\frac{\Delta p_c}{\Delta \varepsilon_c} \right] \qquad [1.5]$$

If true strain is used to calculate $\Delta \varepsilon_c$ then the additional term is not required and [1.4] is correct.

1.12.4 Shear strain

For undrained loadings where the material is incompressible the shear strain at the cavity wall γ_c is given by the constant area ratio, $\Delta A/A$, which is the change of area divided by the current area. Referring to Fig. 1.12 where the cavity starts from unit radius:

$$\frac{\Delta A}{A} = \frac{\pi(1 + \varepsilon_c)^2 - \pi}{\pi(1 + \varepsilon_c)^2} \qquad [1.6]$$
$$\therefore \gamma_c = 1 - 1/(1 + \varepsilon_c)^2$$

Conveniently, γ_c can be expressed in terms of simple strain. Note also that when ε_c is less than 0.01 (1%), then $\gamma_c \approx 2\varepsilon_c$. The error is less than 2%.

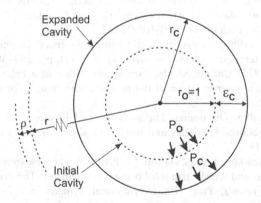

Figure 1.12 Stresses and strain around expanding cavity.

For material that is compressing or dilating, shear strain is more complex and depends on additional calculations that provide the increments of volumetric strain. These are explained in chapter 5.

NOTES

1 The inevitable exception is the Marchetti flat-blade dilatometer, an ingenious device that provides one point of pressure and displacement but only in soils.
2 The converse does not apply. Knowing only the stress required to strain the material 100%, it is not possible to calculate the yield condition. This is the limitation of the cone penetrometer test.
3 Even if the measurement is change of volume, the analysis is using the ratio of two radii quantifying the elastic and plastic extent.

REFERENCES

Bishop, R.F., Hill, R. & Mott, N.F. (1945) The theory of indentation and hardness tests. *Proceedings of the Physical Society* 57 (3), pp. 147–159.

Bolton, M.D. & Whittle, R.W. (1999) A non-linear elastic/perfectly plastic analysis for plane strain undrained expansion tests. *Géotechnique* 49 (1), pp. 133–141.

Carter, J.P., Booker, J.R. & Yeung, S.K. (1986) Cavity expansion in cohesive frictional soil. *Géotechnique* 36 (3), pp. 349–358.

Gibson, R.E. & Anderson, W.F. (1961) In situ measurement of soil properties with the pressuremeter. *Civil Engineering and Public Works Review* 56, 658 (May), pp. 615–618.

Houlsby, G. & Withers, N.J. (1988) Analysis of the Cone Pressuremeter Test in Clay. *Géotechnique* 38 (4), pp. 573–587.

Hughes, J.M.O., Wroth, C.P. & Windle, D. (1977) Pressuremeter tests in sands. *Géotechnique* 27 (4), pp. 455–477.

Mair, R.J. & Wood, D.M. (1987) Pressuremeter Testing. Methods and Interpretation. Construction Industry Research and Information Association Project 335. Publ. Butterworths, London. ISBN 0-408-02434-8.

Manassero, M. (1989) Stress-Strain Relationships from Drained Self Boring Pressuremeter Tests in Sand. *Géotechnique* 39 (2), pp. 293–307.

Palmer, A.C. (1972) Undrained plane-strain expansion of a cylindrical cavity in clay: a simple interpretation of the pressuremeter test. *Géotechnique* 22 (3), pp. 451–457.

Whittle, R.W. (1993) Discussing 'The assessment of in situ stress and stiffness at seven over-consolidated clay and weak rock sites'. *Ground Engineering*, Sept 1993, pp. 19–20.

Whittle, R.W., Hawkins, P.G. & Dalton, J.C.P. (1995) The view from the other side – lift-off stress and the six arm self-boring pressuremeter. *The Pressuremeter and its New Avenues*, Proc. 4th International Symposium 17–19 May 1995, pp. 379–386.

Yu, H.S. & Houlsby, G. (1991) Finite cavity expansion in dilatant soils: loading analysis. *Géotechnique* **41** (2), pp. 173–183.

Yu, H.S. & Houlsby, G. (1995) Large strain analytical solution for cavity contraction in dilatant soils. *International Journal of Numerical & Analytical Methods in Geomechanics* **19**, pp. 793–811.

Yu, H.S. (1996) Interpretation of pressuremeter unloading tests in sands. *Géotechnique* **46** (1), pp. 17–31.

Chapter 2

Determining Modulus

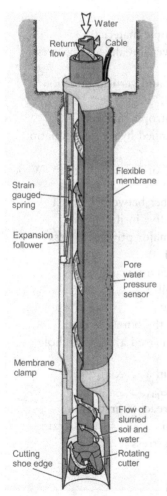

Water

Return flow

Cable

Flexible membrane

Strain gauged spring

Expansion follower

Pore water pressure sensor

Membrane clamp

Flow of slurried soil and water

Cutting shoe edge

Rotating cutter

Figure 2.0 Schematic of the self-boring probe.

DOI: 10.1201/9781003200680-2

2.1 NOTATION

A	General purpose constant term
c_u	Undrained shear strength
G	Shear modulus
G_{ho}	Shear modulus at the insitu stress state
G_i	Initial shear modulus
G_S	Secant shear modulus
G_t	Tangential shear modulus
G_y	Secant shear modulus when the material first reaches full plasticity (yield)
G_{max}	Elastic shear modulus
G_{hh}, G_{vh}	Shear moduli for transversely isotropic material
G_{ref}	Used as the constant term in a power law describing stress dependency
E_h, E_v	Young's modulus in the horizontal and vertical direction
v	Poisson's ratio
v_{hh}, v_{hv}	Poisson's ratios for transversely isotropic material
m	Exponent of curvature used in modified hyperbolic equation
N	Exponent (Whittle & Liu, 2013)
n	Stress level exponent (Janbu, 1963) or anisotropy ratio (Wroth et al., 1979)
z	Proportion of strength so is a number between 0 and 1
k_o	Ratio of horizontal to vertical effective insitu stress
k_a	Ratio of minor principal stress to major principal stress
K	Bulk modulus (also can be denoted B)
τ, τ_c	Shear stress, suffix c means at the cavity wall
p_c	Pressure measured at the cavity wall
p'_o	Initial effective cavity reference stress
ε_c	Circumferential strain measured at the borehole wall
γ	Shear strain, γ_c is shear strain measured at the borehole wall
γ_a	Invariant shear strain (axial strain in a triaxial test)
γ_{ref}	Reference shear strain when G_s is $G_{max}/2$
γ_y	Shear strain at which the material reaches first failure
α	Shear stress constant when the strain scale is shear strain
α_{ref}	Shear stress constant at a reference stress level
β	Exponent of non-linearity
σ'_{av}	Mean effective stress
$\sigma_{ho}\ \sigma'_{ho}$	Total and effective insitu lateral stress

σ_{vo} σ'_{vo} Total and effective insitu vertical stress
ϕ' Angle of shearing resistance

2.2 BACKGROUND

It is a paradox that within a destructive test there is a process which allows straightforward and repeatable measurements to be made of the elastic properties of the ground. That is what is obtained from incorporating small cycles of unloading and reloading into the test procedure and for many users this one aspect alone justifies specifying the high-resolution pressuremeter test.

For lower-resolution pressuremeters the slope of the virgin loading curve has been the primary source of stiffness data. This part of the test (Fig. 2.1, the pseudo-elastic slope) is dominated by the disturbance caused by the method used to place the probe in the ground. Nevertheless, it is still used by empirical practices based on Ménard techniques for pre-bored pressuremeters. Because the material has been completely unloaded prior to the test commencing, the first loading, potentially is an analogue for certain construction processes. However, it is not a fundamental engineering property and there are better ways of estimating the consequences of unloading the ground.

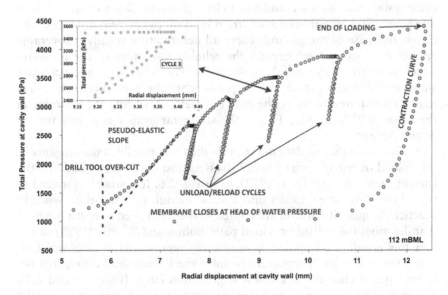

Figure 2.1 Field curve—pre-bored test in silty sand.

Only if the material is rock, or the pressuremeter has been installed by self-boring will the initial slope give stiffness values comparable with those of undisturbed material. Fortunately it is not necessary to use the initial data at all. Once the material is loaded past its yield condition, a zone of plastically deformed material starting from the cavity wall begins to extend into the soil mass. One consequence is that the stress history of the insertion process is erased when all elements within the plastic zone are put into a uniform condition. At a radius remote from the cavity wall and the pressuremeter there will be a single boundary where the material is on the point of yielding. If the direction of loading is reversed, the response seen at the pressuremeter will be that of the material beyond the boundary being unloaded elastically and ultimately plastically if the unloading continues past the point of reverse plastic yield.

In Fig. 2.1, small cycles of unloading and reloading exploit the response prior to the reverse yield condition to derive the elastic properties. It is evident that although each cycle is taken at a different cavity radius and stress, the procedure is highly repeatable. This test was pre-bored, so the cavity was completely unloaded prior to the pressuremeter test commencing. The consequences of this are clear when the slope of the initial loading is compared to the unload/reload cycles.

It would be straightforward to place a line through each of the cycles and use the slope to calculate the shear modulus, G. This is common practice but, unless the material is intact rock, is misleading. The third cycle is shown as an inset, and has a clear hysteretic characteristic. This is due primarily to the influence of strain level on the current modulus. The elastic response of the ground where all deformation is fully recoverable applies to a strain level beyond the reliable resolution of the pressuremeter at shear strains of about 0.001%. The unload/reload cycles are showing the largely recoverable response when the stress alteration is less than that required to make the material yield, this being shear strains in the range 0.01% to 1%. This is the strain range that is significant for design purposes.

If there are sufficient data in the cycle then it is possible take tangents to the unload or reload path of radial stress against cavity strain and find the current shear stress (Palmer, 1972), as in Fig. 2.6. It is straight-forward to turn these shear stress values into a shear modulus degradation trend. In practice the quantity of data are limited so there is a need to find a function that describes the unload or reload path. Bolton and Whittle (1999) shows that this non-linear response is adequately represented by a power law. Using the power law parameters to solve the Palmer semi-differential solution gives a continuous stiffness degradation curve (Figs. 2.2 and 2.3). The individual points in these figures are tangents to the measured data; the lines are the power law trend.

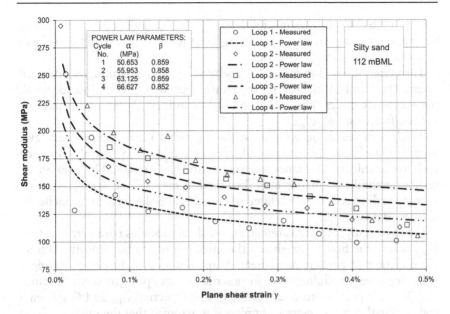

Figure 2.2 Shear modulus degradation curves (drained loading).

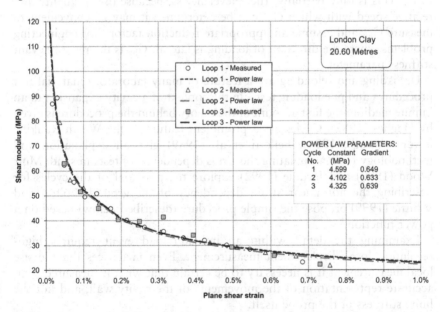

Figure 2.3 Shear modulus degradation curves (undrained loading).

Alternatives to the power law method include Jardine (1992) where a "transformed strain" approach is applied to unload/reload data. These semi-empirical formulations were developed for specific soil types and are not transferable.

If the material is low permeability clay, giving an undrained loading, then the mean effective stress σ'_{av} following yield is constant, and all cycles will follow a similar path (Fig. 2.3).

The material tested in Fig. 2.1 is a silty sand and the loading is a drained event. The trend of each cycle (the strain dependency) is almost identical, but successive cycles (Fig. 2.2) plot a higher stiffness, because of increasing mean effective stress, σ'_{av}.

A full data reduction will adjust these trends to a reference stress level such as the effective *in situ* lateral stress, σ'_{ho}. This requires σ'_{av} for each cycle to be calculated. Hence, although stiffness is obtainable from all insertion methods, no matter how disturbed, it may still be necessary to determine additional strength related parameters in order fully to reconcile the stiffness data.

Used vertically, the pressuremeter gives shear modulus parameters of type G_{hh}, where the first suffix shows the direction of loading and the second the direction of particle movement. Many design calculations requiring a value for shear modulus mean in practice the independent shear modulus G_{vh}. It is not possible to discover the ratio connecting G_{hh} and G_{vh} from a conventional pressuremeter test unless it is assumed that the ratio is related to k_o. This is only partially true. Nevertheless, because of the quality and relative speed with which G_{hh} can be determined it may be convenient to measure G_{hh} and assume an appropriate reduction factor. For engineering problems where the direction of loading is lateral, G_{hh} is the most relevant stiffness parameter.

Unloading and reloading are a feature of many laboratory material test procedures and pile loading tests. In the context of a cavity expansion in an infinite medium the first account of the theory behind the procedure is given by Hughes (1982). Cycles are a prominent feature of the Wroth Rankine lecture (Wroth, 1984). Bellotti et al. (1989) give an explanation and methodology for manipulating the stress dependency of tests in sand. Muir Wood (1990) and Jardine (1992) explore the potential of the cycles for describing the non-linear strain dependency of the ground. Bolton and Whittle (1999) propose the simple procedure (described below) based on a power function.

Examining the detail within an unload/reload event requires high-resolution local displacement measurement. Even in devices that do use local measurement it is necessary to be certain that what is measured is an accurate representation of the movements of the cavity wall, and not the finite stiffness of the probe itself.

The cavity expansion test shears the material. The modulus measured is shear modulus G and is independent of Poisson's ratio v. G can be used to derive Young's modulus E but v must be given or estimated. It is also straightforward to derive the bulk modulus K from the shear modulus.

2.3 DESCRIBING THE UNLOAD/RELOAD CYCLE

Fig. 2.4 shows an unload/reload cycle extracted from a field curve such as the example in Fig. 2.1.

In this example, the data are part of a test in stiff clay. In the interval between pausing the loading to take the cycle and the actual reversal of stress there are several data points showing the expansion continuing for no pressure increase. This phase of time-dependent deformation is a combination of several factors but primarily it is a rate effect. Separating the contribution of the processes that contribute to this behaviour is complex. It is unlikely that it will be possible to wait long enough for all creep behaviour to cease. This test in clay is an undrained event. There is a large excess pore pressure in the soil mass at the commencement of the cycle and waiting more than a short time would allow consolidation to take place. The reducing displacement between readings indicate when creep has fallen to a level low enough to permit a cycle to be taken, and pressure begins to be vented from the system.

The decrease of pressure continues until sufficient data have been recorded to give a clear indication of the path of the unloading response. The pressure decrease needs to be less than that required to cause reverse plastic failure, which for any soil is twice the current mobilised shear stress. In practice the shear stress is not known, but it generally safe to decrease the radial stress at the cavity wall by 25%.

The reloading phase mirrors the unloading with a similar rate of pressure change and eventually crosses the cycle unload path. There are a few points

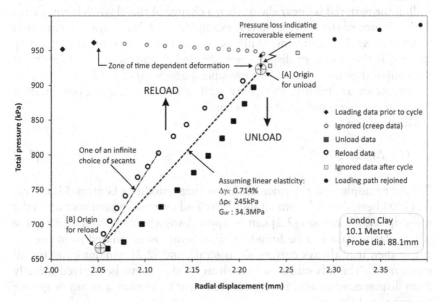

Figure 2.4 Annotated unload/reload cycle.

before the main loading path is re-joined because the cycle only approximately is a recoverable event.

A chord has been drawn through the start and lowest point of the cycle. The slope of this can be used as a means of calculating shear modulus. If the material response was linear elastic, then the result would be the shear modulus. It is apparent that the unloading and reloading data show a non-linear response, and the chord that has been drawn is merely the minimum secant. There are an infinite number of steeper secants that could be drawn. The unloading and reloading responses mirror each other and a rotation of the unloading data would describe the same path as the reloading data. This is made explicit in Fig. 2.6.

2.4 LINEAR ELASTIC INTERPRETATION

If the material response is linear elastic then the local gradient prior to yield can be used as a source of modulus data. In Fig. 2.1, the part of the curve labelled "pseudo-elastic" is an example. Because it occurs at the start of the test it is sometimes referred to as the initial or first loading shear modulus,[1] G_i.

Shear modulus is the quotient of the change of shear stress τ and change of shear strain γ:

$$G = \Delta\tau/\Delta\gamma \qquad\qquad [2.1]$$

Shear stress and shear strain are not directly measured by the pressuremeter. In Fig. 2.4, the axes are radial displacement and radial stress at the cavity wall. If the material is linear elastic then a change of radial stress is equivalent to the change of shear stress (in this example, 244 kPa). Displacement needs to be expressed non-dimensionally as strain. The change of current cavity strain ε_c is the change of displacement divided by the radius of the probe at the midpoint of the cycle, in this example 3.566×10^{-3}.

Current shear strain at the cavity wall γ_c is twice ε_c, an approximation that is convenient and valid if ε_c is small.

This leads to

$$G = \Delta p_c/2\Delta\varepsilon_c \qquad\qquad [2.2]$$

For the example test, the calculation for shear modulus G gives 34.3 MPa. As G has been derived from an unload/reload event it sometimes carries the subscript G_{ur}. Equation [2.2] can be applied anywhere on the loading curve where elastic data can be found. If shear strain is calculated as in the example then it is always current shear strain and [2.2] remains valid for all expansions. The advantage is that shear modulus can be derived directly from displacements, avoiding the requirement to identify a strain origin for the entire field curve.

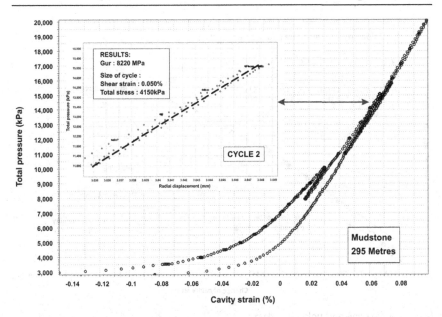

Figure 2.5 A linear elastic unload/reload cycle.

In practice, the only material likely to give a linear elastic response for an extended stress change is intact rock (Fig. 2.5). There are three such cycles in the test all giving similar results (approximately 8 GPa). Because the material is strong enough to support an open hole without failing in the reverse sense, the slope from the latter part of the field curve gives a result similar to the modulus determined from unload/reload data.

2.5 NON-LINEAR STIFFNESS/STRAIN RESPONSE

In all soils, for shear strains up to the yield value, the stiffness/strain relationship is not linear. The unload/reload cycle can be made to give an almost complete description of this relationship by looking at smaller steps of pressure/strain other than the points at the start and finish of the cycle. Fig. 2.6 plots the unloading and reloading data from Fig. 2.4 in this way. Each path has its own origin, as indicated in Fig. 2.4, and the unloading data have been rotated to emphasise that both sets of data are showing the same thing. It follows that it is only necessary to examine one-half of the rebound cycle, and the origin for data obtained after the reversal of stress in a loop has the smaller uncertainty because creep is at a minimum (Whittle et al., 1992). Fig. 2.6 also shows the underlying shear stress response. The test is an undrained event so taking tangents to the radial stress data gives the current shear stress (Palmer 1972, Hughes 1973).

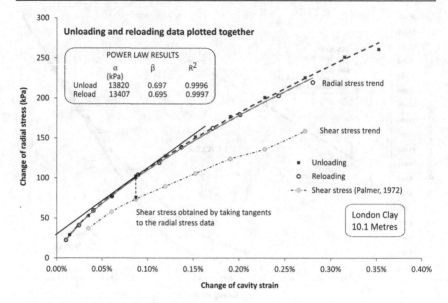

Figure 2.6 Loading and unloading data.

The simplest description of the reloading response is a power law. The exponent of the power law defines the non-linearity of the response and is denoted β. It is generally a number between 0.5 and 1, where 1 indicates linear elasticity. In the example β is almost 0.7, appropriate for a silty clay.

The results in Fig. 2.6 show the power law trend in radial stress/cavity strain space. This is a precursor to the Bolton and Whittle method, where the strain scale is plane shear strain. This starting place is chosen because curve modelling solutions presented in later chapters require stiffness in the form of a radial stress/cavity strain relationship. Using parameters that the pressuremeter test can determine, the connection between the two can be written as:

$$p_c = A\varepsilon_c^\beta \qquad [2.3]$$

"A" is the radial stress constant when the strain scale is current cavity strain (circumferential strain at the cavity wall). Because the test is undrained, there are no volumetric strains so assuming small strains, the shear strain is twice the circumferential strain. The result in radial stress/shear strain space is given by:

$$p_c = \frac{A}{2^\beta}\gamma_c^\beta \qquad [2.4]$$

For an undrained expansion, Palmer (1972) shows that the current shear stress τ_c, is given by:

$$\tau_c = \frac{dp_c}{d[\log_e(\gamma_c)]} \qquad [2.5]$$

Substituting for dp_c using the right-hand side of [2.4] allows [2.5] to be solved giving:

$$\tau_c = \frac{A\beta}{2^\beta}\gamma_c^\beta \qquad [2.6]$$

Bolton and Whittle refer to $\frac{A\beta}{2^\beta}$ as the shear stress constant and call it α. The secant shear modulus, G_s, is the derivative of the shear stress and is given by:

$$G_s = \alpha\gamma_c^{\beta-1} \qquad [2.7]$$

The tangential shear modulus, G_t, for a shear strain γ is given by (Muir Wood, 1990):

$$G_t = G_s + \gamma\left[\frac{dG_s}{d\gamma}\right] \qquad [2.8]$$

Hence it follows from [2.7]:

$$G_t = \alpha\beta\gamma_c^{\beta-1} \qquad [2.9]$$

Equation [2.7] gives a means of determining the secant shear modulus for shear strains below the yielding value, down to 10^{-4}. This is the safe resolution limit of the current generation of pressuremeters but is more than the elastic strain at which the stiffness degradation commences.

It is usual to plot the trend between 10^{-4} and 10^{-2} plane shear strain (0.01% to 1%, see Fig. 2.3). Ideally, the large strain limit should be the yield shear strain of the material. A shear strain of 1% is appropriate for many stiff clays, but will be too large for sands and too small for soft clay. The secant shear modulus at yield strain is G_y and is the secant shear stiffness controlling the pressuremeter loading curve after yield has been achieved.

It is not necessary to take cycles of small strain amplitude to obtain small strain stiffness parameters. It is better to make the cycles as large as practicable (subject to the condition that the material is not allowed to fail in extension) to obtain parameters from as wide a strain range as possible.

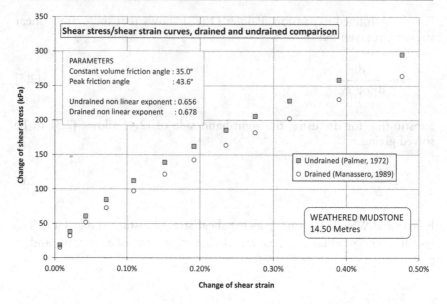

Figure 2.7 Drained and undrained response.

The Bolton and Whittle analysis was developed for undrained tests. For a test in drained material, the solution remains valid if it is assumed that whilst the material is deforming below yield strain there are negligible volumetric strains.

The power approach is merely a curve-fitting exercise and any solution in radial stress/cavity strain space can be used to generate a smooth data set. This allows a numerical solution for drained tests to be applied (Manassero, 1989). The results are similar to the "undrained" parameters with a tendency for β to be slightly higher (more linear). The difficulty is that to apply the drained analysis the ambient water pressure and the constant volume friction angle ϕ'_{cv} must be known or estimated. Fig. 2.7 is an example of a test in highly weathered mudstone, with the shear stress/shear strain response obtained by treating a reloading phase as a drained and undrained event. The difference between the two trends will depend on the potential for dilation.

2.6 STRESS LEVEL

2.6.1 Stress level in soils

For modulus parameters derived from undrained expansion tests, the mean effective stress remains unchanged throughout the expansion and all stiffness/strain data will plot the same trend. Conversely, failure to plot the same trend implies changes in the mean effective stress (Fig. 2.2).

Whittle and Liu (2013) give a method for both stress and strain adjustment and can be applied to tests that contain at least four unload/reload cycles. Their solution can be written as:

$$G_s = A\sigma_{av}^N \tag{2.10}$$

where

$$A = c \ln(\gamma) + d$$
$$N = x \ln(\gamma) + z$$

A and N are semi-log equations incorporating shear strain and the coefficient and constants are found by re-plotting the modulus data to extract the stress and strain dependency. For most purposes, this level of complexity is not required and a simpler approach can be adopted.

1. Start by carrying out the non-linear analysis described previously and discover pairs of α and β values. Use these to find, for each cycle, G_s at an intermediate value of shear strain, such as 0.1%.
2. Calculate the mean effective stress σ'_{av} at the commencement of each loop:

$$\sigma'_{av} = \frac{p'_c}{3}[1 + k_a + \sqrt{k_a}] \tag{2.11}$$

where

$$K_a = \frac{\sigma_1}{\sigma_3} = \frac{1 - \sin\phi'}{1 + \sin\phi'}$$

ϕ' is the peak angle of internal friction
p'_c is the effective radial stress at the cavity wall[2]

p'_c is the pressure measured by the pressuremeter minus the ambient pore water pressure and whilst it is increasing σ'_{av} is approximated by [2.11]. This is the mean effective stress at the cavity wall—it has been pointed out by several commentators that $p'_c/2$ is generally a good approximation. Houlsby and Schnaid (1994) show experimentally that the stress effect is dominated by conditions immediately adjacent to the pressuremeter surface.
3. Plot modulus against effective stress (Fig. 2.8).

The example shows three tests in sand with multiple unload/reload cycles treated in this way. Each test gives a set of points that follow a power law

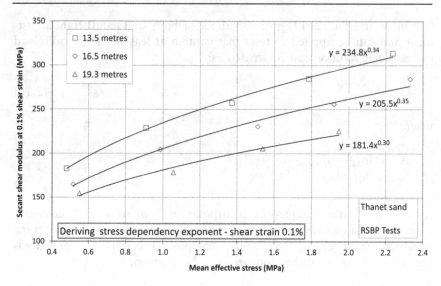

Figure 2.8 Finding the stress dependency exponent.

trend. The exponent of the power law is describing the stress dependency at this level of shear strain. At this strain, typical values for the exponent n are about 0.35 so there is a significant stress dependency. The correlation coefficient for each trend is 0.99.

Given the exponent n, for each cycle a stress adjusted version of α is found, α_{ref}:

$$\alpha_{ref} = \alpha \left[\frac{\sigma'_{ref}}{\sigma'_{av}} \right]^n \qquad\qquad [2.12]$$

Equation [2.12] incorporates the relationship suggested by Janbu (1963) and forms the basis of the approach to stress dependency used in Bellotti et al. (1989). The reference stress is typically σ'_{ho}.

Figs. 2.9 and 2.10 are "before" and "after" examples of the method being applied to a test in dense sand. The scales are similar in both plots, and it is apparent that in practice the stress adjustment gives a trend very similar to that of the first uncorrected cycle. The corrected trend also shows convergence at the strain level used for finding the stress exponent n. This is expected because it is known that the stress adjustment is slightly dependent on strain level. The Whittle and Liu (2013) solution varies n with strain level but for most practical purposes the refinement is not required.

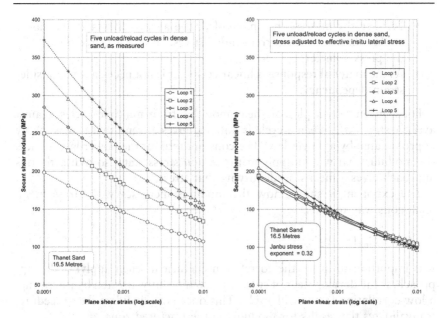

Figure 2.9 Unadjusted stiffness trend.　　*Figure 2.10* Stress adjusted stiffness trend.

2.6.2 Stress level in rocks

In material that has yet to fail in shear and the values for shear modulus from unload/reload cycles suggest linear elastic behaviour, then the stress dependency is straightforward to derive. It is less complex than the procedure for soils. Fig. 2.11 is an example.

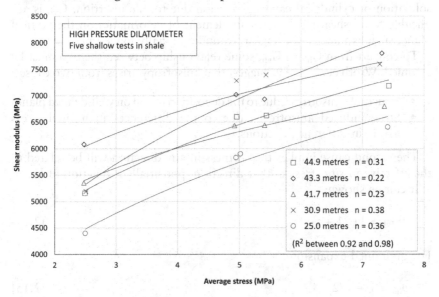

Figure 2.11 Finding the stress dependency exponent in rock.

- Because the material has not failed, the average stress at the cavity wall is about half the current radial stress, and is measured directly by the pressuremeter.
- As the material response is linear elastic, it is not necessary to consider strain dependency.

For a given test, a plot of shear modulus from multiple cycles against average stress, σ_{av}, when a cycle is initiated gives a trend that is reasonably approximated by a power law. The constant term of the power law, G_{ref}, is the modulus that would apply at the *in situ* stress state, so indirectly this approach gives a means of identifying that stress.

The exponent n shows how the modulus increases with stress in an analogous way to soils:

$$G_{ur} = G_{ref}[\sigma'_{av}]^n \qquad [2.13]$$

It is important to quote the correlation coefficient when applying this approach. If the development of stiffness is significantly affected by fracturing, a low correlation factor will apply. This does not invalidate the procedure; it invalidates the results for modulus in the fractured zone.

The test curves in Fig. 2.11 give exponents of stress dependency between 0.2 and 0.4.

2.7 HORIZONTAL-VERTICAL ANISOTROPY

The pressuremeter test gives values for G_{hh}, the shearing stiffness in the horizontal plane. This is directly applicable to the analysis of radial consolidation or cylindrical cavity expansion due to pile insertion. G_{vh} is applicable to all shearing that has an element of deformation in the vertical plane, such as under a footing or around an axially loaded pile.

To convert from G_{hh} to G_{vh}, some relationship between the two must be assumed. Wroth et al. (1979) suggest that anisotropy arises from two causes:

- Structural anisotropy, due to the deposition of soil on well-defined planes
- Stress-induced anisotropy, due to the differences in normal stress acting in different directions.

The second cause implies the stiffness in any direction will be related to the effective *in situ* stress in that direction, potentially a function of k_o.

It can be shown:

$$2G_{hh} = E_h/(1 + \upsilon_{hh}) \qquad [2.14]$$

For undrained expansion

$$\upsilon_{hh} = 1 - n/2 \qquad [2.15]$$

and

$$n = E_h/E_v \approx k_0 \qquad [2.16]$$

From this it follows:

$$E_h = (4 - n)G_{hh} \qquad [2.17]$$

and

$$E_v = (4 - n)G_{hh}/n \qquad [2.18]$$

This is as far as an argument from first principles can be taken. k_0 for most material lies between 0.8 and 2; hence, n is unlikely to be significantly less than 1. From [2.16] and [2.17], E_h/G_{hh} lies between 2 and 3. From [2.18], E_v/G_{hh} lies between 1 and 3.

It is reasonable to suppose that G_{vh} will be linked to E_v by Poisson's ratio in a relationship of the form of [2.14]. Plausible values of E_v/G_{vh} would seem to be 2.4 to 3, so in a material with k_0 of 2, G_{vh} could be as low as $G_{hh}/3$. Simpson et al. (1996) came to the same conclusion, but find in practice heavily over-consolidated London clay approximates to $G_{vh} \approx 0.65G_{hh}$. The influence of the strain range is not separately considered in these studies. Fig. 2.14 shows results from London Heathrow Terminal 5 where $G_{vh} \approx 0.62G_{hh}$. Brosse et al. (2017), report ratios higher than this from four distinct stiff clay soils.

Lee and Rowe (1989) give details of the anisotropy characteristics of many clays varying from lightly over-consolidated to heavily over-consolidated. The general conclusion is E_v/G_{vh} lies between 4 and 5, rather more than the isotropic relationship of 3. They were concerned with the impact of anisotropic stiffness properties on surface settlement so deriving G_{vh} from E_v is unsatisfactory—although G_{vh} is insensitive to the direction of loading, E_v is not.

In all studies, G_{hh} is greater G_{vh}. How much so is not clear, and whether the difference is constant over the entire sub-yield strain range is also not established.

2.8 SHEAR MODULUS FROM OTHER PARTS OF THE PRESSUREMETER CURVE

The first part of the unloading is an elastic process and can be used as a source of stiffness information. By the time the pressuremeter unloads, creep strains due to consolidation and rate effects will be large, so there will be a tendency for the initial unloading to be too stiff. If some allowance is made for this, then reasonable estimates of the shear modulus will be obtained.

Curve-fitting analyses imply a value for the secant shear modulus at yield. Although this is not likely to be the best way of deriving shear modulus data, it is important justification for using such analyses that they can predict this independently measurable stiffness.

2.9 ELASTIC (YOUNG'S) MODULUS, E

All modulus parameters derived from the vertically orientated pressuremeter test are shear modulus, G_{hh}. They can be converted to Young's modulus, E_h, using [2.14].

A non-linear power law based version of [2.14] is

$$E = 2\alpha(1 + v)\left(\sqrt{3}\gamma_a\right)^{\beta-1} \qquad [2.19]$$

where γ_a is invariant or axial strain

$$\gamma_a = \frac{\gamma_c}{\sqrt{3}} \qquad [2.20]$$

Equation [2.20] reconciles the plane shear strain of cavity loading with invariant shear strain used in triaxial apparatus (Muir Wood, 1990), making it easier to compare pressuremeter and laboratory-derived results.

2.10 BULK MODULUS (K OR B)

The bulk modulus is sometimes termed the modulus of compressibility and describes the resistance of a material to uniform hydrostatic pressure. For a given Poisson's ratio, v, it can be derived from the shear modulus, G, as follows:

$$K = \frac{E}{3(1 - 2v)} = \frac{2G(1 + v)}{3(1 - 2v)} \qquad [2.21]$$

K varies with the ratio of axial to compressive strain (Poisson's ratio) and non-linearity in this context is a function of the amplitude of volumetric strain change. The non-linearity described by [2.7] is a non-linear response to changing shear strain. Throughout an unload/reload cycle, the material is deforming at constant volume and, hence, under the condition of a constant bulk modulus. When deriving K parameters from pressuremeter data, it follows that the "G" in [2.21] should be a very small strain value.

2.11 NON-LINEAR MODULUS IN TERMS OF SHEAR STRESS

It is sometimes convenient to derive stiffness values as a proportion of the mobilised strength. If the shear strength, c_u, or shear stress at first failure, τ_f, is known, and z represents the proportion of strength used, then the shear strain, γ_z, for this proportion is given as follows:

Where $0 < z \leq 1$:

$$\gamma_z = \left(\frac{zc_u}{\alpha}\right)^{\frac{1}{\beta}} \qquad [2.22]$$

For example, it is common to require G_{50}, the shear modulus, when half of the available strength is mobilised. It is straightforward to apply the preceding non-linear stiffness expressions to derive the relevant modulus.

Generally, shear modulus at strength fraction z

$$G_{(z)} = \alpha\left(\frac{zc_u}{\alpha}\right)^{\frac{\beta-1}{\beta}} \qquad [2.23]$$

Specifically, for G_{50}:

$$G_{50} = \alpha\left(\frac{c_u}{2\alpha}\right)^{\frac{\beta-1}{\beta}} \qquad [2.24]$$

τ_f can be used in place of c_u for tests in drained materials. It will be approximately $p'_0 \sin\phi'$, where p'_0 is the effective cavity reference pressure and ϕ' is the angle of internal friction.

2.12 POSSIBLE METHODS FOR ESTIMATING G_{MAX} AND THE THRESHOLD ELASTIC SHEAR STRAIN

The Bolton and Whittle (1999) procedure is valid for shear strains in the range 10^{-4} to the material yield strain. The lower limit is due to residual friction in the mechanical elements of the strain measuring devices, masking the threshold strain at which the decay process commences. The full stiffness decay curve is usually described with a modified hyperbolic function. Fig. 2.12 sketches the connection between the undrained power law representation of shear modulus decay and the equivalent hyperbolic curve where the horizontal axis is log normalised shear strain.

The stiffness decay function is the derivative of the stress/strain response, so the area under the tangential decay trend is the undrained shear strength, c_u. The component that is omitted in the Bolton and Whittle analysis is the area marked τ_e. As a quantity, it is vanishingly small and ignoring it has negligible consequences for analysing strength. Nevertheless, a method for including it is required in order to marry the completeness of the hyperbolic function with the precision of the power law decay description.

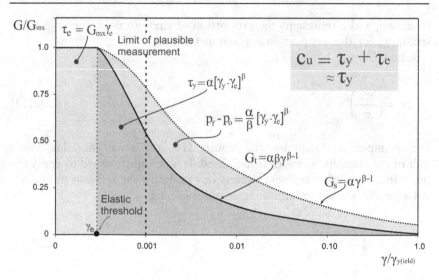

Figure 2.12 Undrained non-linear stiffness using a power law description.

Oztoprak and Bolton (2013) use the following to model the decay curves for a wide range of tests conducted in sand:

$$\left(\frac{G}{G_{max}}\right) = 1/\left[1 + \left(\frac{\gamma - \gamma_e}{\gamma_{ref}}\right)^m\right]$$

[2.25]

where
  G is secant shear modulus at a shear strain γ
  γ_e is the shear strain at the elastic threshold
  γ_{ref} is the reference shear strain when $G/G_o = 0.5$
  The exponent m controls the curvature of decay.

There is nothing specific to sand about [2.25] apart from the curvature parameter m. Oztoprak and Bolton found that $m = 0.88$ (in their paper, this parameter is denoted a, but to avoid confusion, here it is called m) gave the best average fit to their database of sand tests. It is reasonable to suppose that the curvature is related to particle size and shape and consequently will vary with soil type.

For the pressuremeter test with unload/reload cycles, the majority of the stiffness decay relationship is known, but G_{max} is not. This invites an iterative approach where successive estimates of G_{max} can be used in [2.25] to find a modified hyperbolic trend that reproduces the pressuremeter decay data (Fig. 2.13). The horizontal axis is a proportion of mobilised shear stress using [2.23] to relate shear strain to mobilised shear stress. An iterative process is used to re-calculate the reference shear strain, γ_{ref}, and threshold shear strain, γ_e, whenever the estimate of G_{max} is altered.

To produce the trend, the hyperbolic curvature parameter, m, has to be chosen in addition to making estimates of G_{max}. Results for clay suggest m is

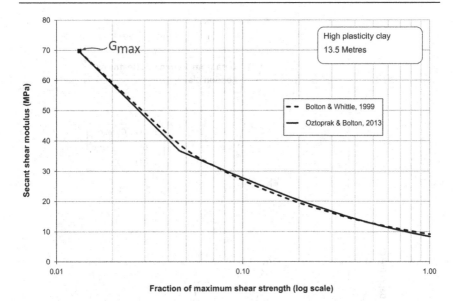

Figure 2.13 Matching hyperbolic and power curves.

about 0.5. This does not agree with Vardanega and Bolton, who report curvature values on average slightly greater than 0.7 but with considerable scatter.

This speculative procedure has been applied to a number of sites and varying materials using different pressuremeter types. The factor m is a number possibly calculable from the non-linear elastic exponent, β. The correlation coefficient between data sets predicted by the power law parameters and the hyperbolic parameters is generally greater than 0.995. There is little judgement involved in setting G_{max} — it is adjusted until the correlation coefficient reaches a peak.

Fig. 2.14 is an example, comparing high-quality laboratory-derived small strain stiffness results (Gasparre et al., 2007) with those derived from self-bored pressuremeter tests at the same site.

Cao et al. (2002), in the context of deriving non-linear elastic solutions for undrained cavity expansion tests, give yield expressions that apply a power law or simple hyperbolic function. By assuming that they are equivalent, it is possible to derive the following:

$$G_{mx} = G_y \cdot \exp\left(\frac{1}{\beta}\right)$$ [2.26]

As it is particularly easy for the power law to calculate G_y (see [2.23] with z = 1), this would seem to give a straightforward means of finding G_{mx}. Experience to date suggests that the results for G_{mx} obtained in this way are conservative. Note that as β is not likely to be smaller than 0.5, $G_{mx} \approx 7.4 G_y$ is the predicted limit.

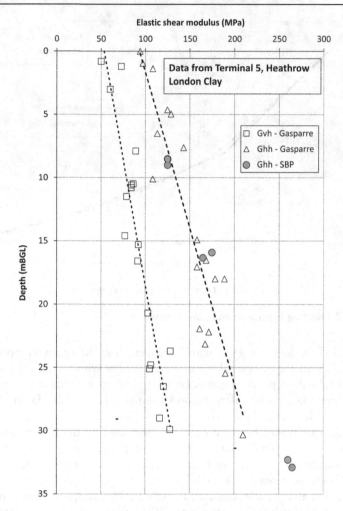

Figure 2.14 Comparing laboratory and pressuremeter results for G_{max}.

2.13 NORMALISATION

The choice of normalisation parameter depends on the underlying purpose. Generally it is the need for a non-dimensional version of a data set where the influence of depth has been removed. In the case of a pressuremeter test, the usual choices are the initial effective stress p'_0 or strength, typically undrained shear strength c_u in the case of data from tests in fine-grained material. A less commonly used though potentially more logical parameter is the effective yield stress, p'_{yield}, which combines both and is linked to the over-consolidation ratio. An example of all of these methods applied to a reasonably continuous profile of pressuremeter tests is given in Fig. 2.15.

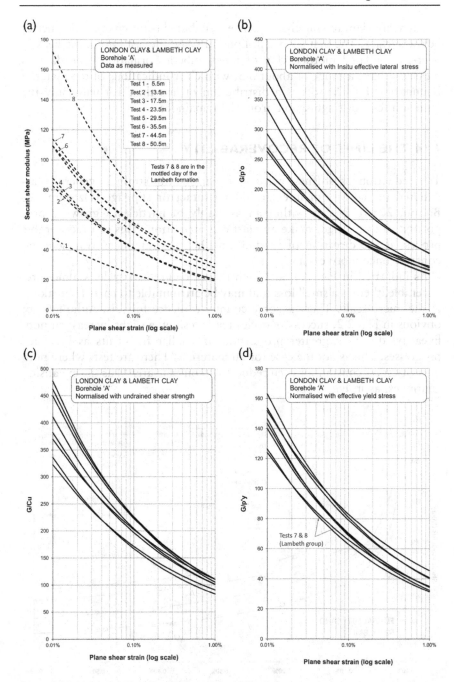

Figure 2.15 Normalising stiffness/strain data (a, b, c, d).

There are eight tests in the sequence distributed between 5 and 51 metres below surface. The first six are in London Clay; the last two in the heavily over-consolidated mottled clay of the Lambeth Formation. The data as measured have a range of about 3.5, with clear indications of the material boundaries. Fig. 2.15d has a distribution of less than 1.3 with only the Lambeth tests distinguishable from the remainder.

2.14 THE LIMIT OF RECOVERABILITY

The reduction of stiffness with increasing shear strain is generally attributed to the loss of inter-granular contact or slippage (Oztoprak & Bolton, 2013) so that within an assembly of particles a growing proportion can no longer make an elastic contribution. This is a recoverable process if the direction of loading and hence straining is reversed and contacts once more engage.

As indicated in Fig. 2.4, although the unload/reload cycle is almost recoverable, there is a small loss that may be attributable to particle breakage at the micro-scale. There is some counter-evidence for this. Although not obvious in Fig. 2.2, successive cycles tend to show a slight increase in non-linearity, due to a greater proportion of smaller fragments as breakage progresses. This is not the case for all materials. There are tests where non-linearity stays constant or even reduces (becomes nearer 1) as the expansion continues, as in Fig. 2.16.

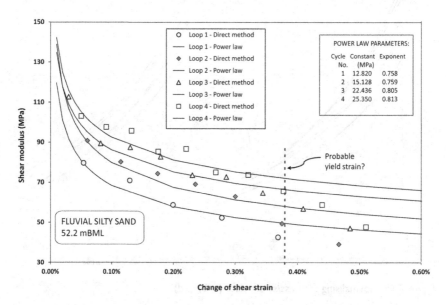

Figure 2.16 Reducing non-linearity in a fluvial sand.

The material in this example has been transported a considerable distance in fast-flowing water, a tumbling process that removes asperities. The more the particles resemble spheres, the less likely it is for microfracturing to occur as there are fewer opportunities for differential stress development.[3]

The opposite condition is extensive crushing of the material, which can make it very difficult to achieve a successful unload/reload cycle. Chalk is an example. Once the yield stress of the chalk has been exceeded, there is a tendency for the structure of the material to collapse. This process will continue indefinitely whilst the loading stress is maintained and is exhibited as a large creep displacement whenever the loading is paused. Reducing the pressure in an attempt to drop below the current yield surface will often result in cycles that are "V" formed (Fig. 2.17).

The chord through the cycle has been aligned with the reloading data, which in this case is the conservative option. The only cycle in this test to be relatively free of creep influence is the one on the final unloading.

Difficulty with creep is also a characteristic of pushed tests, where the radial stress at the time of initiating a cycle is approximately the limiting condition.

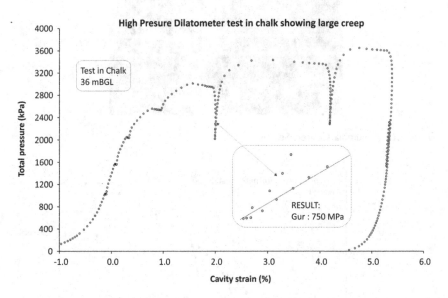

Figure 2.17 The effects of severe creep.

2.15 COLUMBIA CENTER, SEATTLE

Columbia Center is the tallest skyscaper in Washington State. The initial excavation incorporated a tieback shoring wall of unusual height that was beyond the limits of standard design calculations. A finite element model was developed and stiffness values from pressuremeter tests carried out by the first author were provided to initialise the model. It is one of the earliest examples of unload/reload pressuremeter data being used for this purpose (Figs. 2.18 and 2.19).

Figure 2.18 Columbia Center, Seattle.

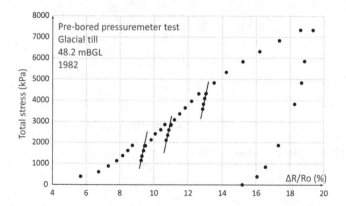

Figure 2.19 Columbia Center test example.

The work was carried out in 1982 using analogue output versions of the modern probes, with readings written down by hand. This is why there are so few data points, especially in the unload/reload cycles which have been treated as linear elastic events. Nevertheless, despite the relative lack of sophistication of both analysis and measurement, the raw unmodified results were found to give good agreement (within a factor of 2) with the measured deformations of the completed structure (Grant & Hughes, 1986).

It is interesting to note that the agreement was better for deformations above the base of the excavation, and the authors observe that higher modulus values were required to match the sub-base response. If a non-linear interpretation of the cycles had been available then this requirement could have been met.

Some parts of this chapter may seem complex. As this example illustrates, the crucial aspect is to have a representative value for shear modulus and an approximate understanding of the strain level to which it applies. This information is available without any difficult interpretation within minutes of completing a test, and this has been the case for a very considerable time.

2.16 SIGNIFICANT MODULUS RELATIONSHIPS SUMMARISED

Shear modulus, G, where τ is shear stress and γ is shear strain:

$$G = \tau/\gamma \qquad [a]$$

G in terms of cavity strain, ε_c, and cavity pressure, p_c:

$$2G = \delta p_c / \delta \varepsilon_c \qquad [b]$$

This is valid for a linear elastic response and a small strain alteration.

Elastic (Young's) modulus, E, in terms of G, where v is Poisson's ratio:

$$E = 2(1 + v)G \qquad [c]$$

Non-linear secant shear modulus G_s:

$$G_s = \alpha\gamma^{\beta-1} \qquad [d]$$

where α is the shear stress constant and β is the exponent of linearity obtained from fitting the reloading response in shear stress/shear strain space with a power function. γ is plane shear strain.

Non-linear secant Young's modulus, E'_s, using invariant shear strain, γ_α:

$$E'_s = 2\alpha[1 + \nu]\left[\sqrt{3}\gamma_\alpha\right]^{\beta-1} \qquad [e]$$

Multiplying by $\sqrt{3}$ converts γ_α to γ, assuming no volumetric strains are involved.

Non-linear tangential shear modulus, G_t:

$$G_t = \alpha\beta\gamma^{\beta-1} \qquad [f]$$

Plane shear strain at failure, γ_f, undrained case, c_u, is undrained shear strength:

$$\gamma_f = [c_u/\alpha]^{1/\beta} \qquad [g]$$

Secant shear modulus at failure, G_y, in terms of stress:

$$G_y = \alpha[c_u/\alpha]^{(\beta-1)/\beta} \qquad [h]$$

For secant shear modulus at any proportion of mobilised stress up to failure, introduce n where $0 < n \leq 1$:

$$G_n = \alpha[nc_u/\alpha]^{(\beta-1)/\beta} \qquad [i]$$

For the particular case of G_{50} at half of the ultimate shear strength:

$$G_{50} = \alpha[c_u/2\alpha]^{(\beta-1)/\beta} \qquad [j]$$

Substitute yield stress τ_f for c_u in the case of drained tests when using these equations.

Speculative calculation for the elastic shear modulus, G_{max}:

$$G_{max} = G_y \, Exp[1/\beta] \qquad [k]$$

NOTES

1 Unfortunately, "initial" is used in two conflicting senses. In the context of non-linear stiffness "initial modulus" generally refers to the maximum or elastic modulus, which in the case of the test in Fig. 2.1 is certainly not true.

2 Note: Eventually when unloading the cavity the radial stress becomes the minor principal stress so k_a in [2.12] is replaced by $1/k_a$. However, for most of the unloading, the mean effective stress will be the state that applied at the end of the loading.

3 The results obtained by applying sub-tangents directly to the measured data show modulus falling below the power law trend at larger shear strain. The strain scale extent is arbitrary; what this difference is indicating is the shear strain beyond which the available shear stress is fully mobilised, so potentially the yield strain of the material.

REFERENCES

Bellotti, R., Ghionna, V., Jamiolkowski, M., Robertson, P. & Peterson, R. (1989) Interpretation of moduli from self-boring pressuremeter tests in sand. *Géotechnique* 39 (2), pp. 269–292.

Bolton, M.D. & Whittle, R.W. (1999) A non-linear elastic/perfectly plastic analysis for plane strain undrained expansion tests. *Géotechnique* 49 (1), pp. 133–141.

Brosse, A., Hosseini Kamal, R., Jardine, R.J. & Coop, M.R. (2017) The shear stiffness characteristics of four Eocene-to-Jurassic UK stiff clays. *Géotechnique* 67 (3), pp. 242–259.

Cao, L.F., Teh, C.I. & Chang, M.F. (2002) Analysis of undrained cavity expansion in elasto-plastic soils with non-linear elasticity. *International Journal for Numerical and Analytical Methods in Geomechanics* 26, pp. 25–52.

Gasparre, A., Nishimura, S., Minh, N.A., Coop, M.R. & Jardine, R.J. (2007) The stiffness of natural London Clay. *Géotechnique* 57 (1), pp. 33–47.

Grant, W. P., Hughes, J.M.O. (1986) Pressuremeter tests and shoring wall design. Proc. of In Situ '86, Virginia Tech., Blacksburg, Virginia, June 23–25, 1986. Publ. ASCE, New York. pp. 588–601 ISBN 0-87262-541-9

Houlsby, G.T. & Schnaid, F. (1994) Interpretation of shear moduli from cone pressuremeter tests in sand. *Géotechnique* 44 (1), pp. 147–164.

Hughes, J.M.O. (1973) An instrument for in situ measurement in soft clays, PhD Thesis, University of Cambridge.

Hughes, J.M.O. (1982) Interpretation of pressuremeter tests for the determination of elastic shear modulus. *Proc. Engng Fdn Conf. Updating Subsurface Sampling of Soils and Rocks and Their in-situ Testing*, Santa Barbara, pp. 279–289.

Janbu, N. (1963) Soil compressibility as determined by oedometer and triaxial tests. *Proc. 3rd Eur. Conf. Soil Mech.*, Wiesbaden, 2, pp. 19–24.

Jardine, R.J. (1992) Nonlinear stiffness parameters from undrained pressuremeter tests. *Canadian Geotechnical Journal* 29, pp. 436–447.

Lee, K.M. & Rowe, R.K. (1989) Deformations caused by surface loading and tunnelling: the role of elastic anisotropy. *Géotechnique* 39 (1), pp. 125–140.

Manassero, M. (1989) Stress-strain relationships from drained self boring pressuremeter tests in sand. *Géotechnique* 39 (2), pp. 293–307.

Muir Wood, D. (1990) Strain dependent soil moduli and pressuremeter tests. *Géotechnique* 40 (3), pp. 509–512.

Oztoprak, S. & Bolton, M.D. (2013) Stiffness of sands through a laboratory test database. *Géotechnique* 63 (1), pp. 54–70.

Palmer, A.C. (1972) Undrained plane-strain expansion of a cylindrical cavity in clay: a simple interpretation of the pressuremeter test. *Géotechnique* **22** (3), pp. 451–457.

Simpson, B., Atkinson, J.H. & Jovicic, V. (1996) The influence of anisotropy on calculations of ground settlements above tunnels. *Proc. Int. Symp. Geotechnical Aspects of Underground Construction in Soft Ground*, City University, London, April.

Vardanega, P.J., & Bolton, M.D. (2013) Stiffness of clays and silts: normalizing shear modulus and shear strain. *Journal of Geotechnical and Geoenvironmental Engineering* **139**, pp. 1575–1589.

Whittle, R.W. (1999) Using non-linear elasticity to obtain the engineering properties of clay. *Ground Engineering* May, **32** (5), pp. 30–34.

Whittle, R.W., Dalton, J.C.P. & Hawkins, P.G. (1992) Shear modulus and strain excursion in the pressuremeter test. *Proc. Wroth Memorial Sym.* Oxford, July 1992.

Whittle, R.W. & Liu, L. (2013) A method for describing the stress and strain dependency of stiffness in sand. *Proc. Symp. ISP6*, Paris – September 4, 2013, session 3, paper 7.

Wroth, C.P. (1984) The interpretation of in situ soil tests. Twenty Fourth Rankine Lecture. *Géotechnique* **34** (4), pp. 449–489.

Wroth, C.P., Randolph, M.F., Houlsby, G.T. & Fahey, M. (1979) A review of the engineering properties of soils with particular reference to the shear modulus. Cambridge University Engineering Department, Soils TR75.

Chapter 3

Analyses for *in situ* Lateral Stress

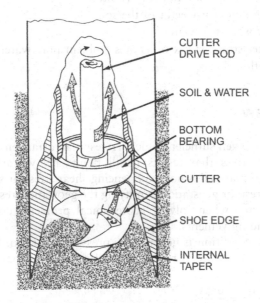

CUTTER
DRIVE ROD

SOIL & WATER

BOTTOM
BEARING

CUTTER

SHOE EDGE

INTERNAL
TAPER

Figure 3.0 Self-boring pressuremeter cutter assembly.

3.1 NOTATION

p_o	Initial cavity reference stress (total)
p_f	Pressure when yield first occurs at the cavity wall (total)
σ_{ho}	Total in situ lateral stress
σ_{vo}	Total in situ vertical stress
σ'_{ho}	Effective in situ lateral stress
σ'_{vo}	Effective in situ vertical stress
σ_H, σ_h	Major and minor total in situ lateral stress (rock)

DOI: 10.1201/9781003200680-3

k_o	Ratio of horizontal to vertical effective in situ stress
r_o	Initial cavity radius
p_1, p_2, p_3	Three readings of stress at 120° from each other
a, b	Minor and major stress
c_u	Undrained shear strength
c'	Drained cohesion
ϕ'	Angle of internal friction
τ_f	Shear stress when yield first occurs
G_i	Shear modulus from the initial slope of the loading curve
α	Shear stress constant (Bolton & Whittle, 1999)
β	Exponent of non-linearity (Bolton & Whittle, 1999)
γ_s	Saturated unit weight of the soil
γ_w	Unit weight of water
u	Water stress, suffix o denotes ambient pore water pressure
z	Depth

3.2 OVERVIEW

Natural ground, given sufficient time, arrives at an arrangement of vertical and horizontal stress that results in all points below surface being in equilibrium. The material is not experiencing shear stress or shear strain. It is said to be at rest, or geostatic. Identifying the geostatic stress components is a necessary step when solving geotechnical problems concerned with assessing ground movements.

The geostatic condition is generally expressed as a ratio, k_o, the coefficient of the earth pressure at rest. It is an effective stress parameter:

$$k_o = \frac{\sigma_{ho} - u_0}{\sigma_{vo} - u_0} = \frac{\sigma'_{ho}}{\sigma'_{vo}} \qquad [3.1]$$

where
   σ_{ho} is the total *in situ* horizontal stress
   σ_{vo} is the total *in situ* vertical stress
   u_0 is ambient pore water pressure

σ_{vo} is straightforward to estimate from the product of the unit weight of the material and depth. σ_{ho} is difficult to estimate reliably, especially so if the material is over-consolidated. It has to be determined experimentally, typically through displacement measurement of the material response to a lateral load or removal of such a load.

Pressuremeters do not measure k_o, but the test can be used to assess σ_{ho}. As all other unknowns are easily derived, it becomes possible to give an estimate of k_o.

The requirement to identify the initial lateral stress is also a problem for the analytical interpretation of a cavity expansion test. Solutions that extract the stress/strain properties of the ground from the pressure/displacement data also require knowledge of the reference condition. In this context, the initial stress is more usually designated cavity reference pressure, p_o, which may be synonymous with σ_{ho}. The initial radius, r_o in Fig. 3.1, is sometimes more easily recognised by a change of gradient in the loading path.

Although in principle the stress ordinate of the initial radius ought to be p_o, in practice this cannot be the case unless the probe has been inserted into the material with insignificant disturbance. However, assuming the two values are co-incident is a first approximation that later examination may revise.

For insertion methods that imply stress *relief*, the origin is taken to be the point where the cavity, which has moved inwards, is restored to its original dimensions. A reference stress, p_o, is identified (by various means) and the displacement ordinate of this stress (r_o in Fig. 3.1) is used to convert subsequent displacements to strain.

Fig. 3.1 shows a "perfect" pre-bored test. For a similarly "perfect" self-bored test, the expansion will not commence until the applied stress reaches p_o, allowing the cavity reference stress to be identified by inspection. This is the so-called "lift-off" method.

The self-boring pressuremeter is also fitted with pore pressure transducers. The trend of excess pore water pressure against total stress can be used to identify yield stress and reference stress.

It is also possible to recognise by inspection the shear stress limit (the point marked p_f in Fig. 3.1) as this is indicated by the onset of a markedly non-linear response. An iterative procedure first suggested by Marsland & Randolph (1977) allows p_o to be back-calculated. The published method is not valid for drained tests, or tests in material with non-linear elastic properties. This effectively rules out all soils. Nevertheless, it is common

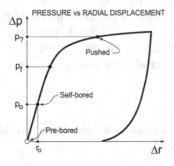

Figure 3.1 Key stress points.

practice to apply the procedure because it tends to set an upper limit to any estimate of reference stress.

Elements of these methods are outlined in Mair & Wood (1987). For a pushed test, the expansion commences when the material is already in an advanced plastic state, $p_?$ in Fig. 3.1. The question mark indicates that although the pressure can be seen, the radial displacement ordinate cannot be used as part of a meaningful strain calculation.

A more novel approach is the balance pressure check (BPC) method (Hoopes & Hughes, 2014). This is a procedure that examines in close detail the latter stages of the final contraction. A small reload/unload event is carried out where the steps of pressure are held constant and the magnitude and direction of any time-dependent movements are examined. What the procedure hopes to identify is a null stress, without discernible movement either inwards or outwards and this "balance pressure" is assumed to be the geostatic lateral stress. Potentially, this gives a means of obtaining reference stress estimates for all types of pressuremeter test, no matter how disruptive the insertion process.

Some rigour can be applied to all these procedures by using the full set of parameters derived from a pressuremeter test within a closed-form model to discover whether the measured field curve can be recovered. The input data set is then adjusted in a controlled manner until the best match for all parameters is obtained. In certain models (Whittle, 1999), the only free parameter is the reference stress.

Other than the BPC method, modelling is the only means of obtaining a reference stress by analytical methods from an undrained pushed test where the cavity expansion commences from zero (Houlsby & Withers, 1988).

In all cases, what is determined is cavity reference pressure, p_o. It is not possible to measure the *in situ* lateral stress σ_{ho} because the act of placing instrumentation always results in some movement, even if very small, and movement means a change in stress. Due to the non-linear nature of soil stiffness a movement of a few micro-strain can result in a large stress alteration. The methods above are indirect indicators for determining σ_{ho}. It is open to question whether p_o is equivalent to σ_{ho}, and multiple sources are examined in order to decide if the assumption is plausible. External evidence might take the form of using the derived reference stress within a k_o calculation, or checking that the derived vertical/horizontal anisotropy can be supported by the material shear strength:

$$\sigma_{ho} - \sigma_{vo} < 2c_u. \qquad [3.2]$$

In practice, there is a wide range of values that would satisfy this condition so its usefulness is limited.

3.3 LIFT-OFF

This method ought to be applicable only to the relatively low-disturbance self-bored test, but occasionally it seems to give sensible results with more extreme insertion procedures. In principle, it is straightforward. The instrument is assumed to be bored into the ground with insignificant disturbance caused to the surrounding material. If the *in situ* conditions around the instrument remain unchanged by the insertion process, then the pressure at which the membrane first moves and the cavity begins to expand is p_o. The corresponding cavity diameter will be the same as the at-rest diameter of the instrument. Because the initial part of a self-bored test is very stiff, the choice is made from an enlarged view of the first 0.5 mm ($\approx 1.2\%$ cavity strain) of the expansion (Fig. 3.2). This range is sufficient to show any elastic behaviour and the early stages of the expansion post-yield.

Difficulties arise because the instrument has finite stiffness and hence there is instrument compliance to be separated from the expansion of the cavity. At the start of the test when the internal pressure is ambient the instrument is being externally loaded by the lateral stress in the ground. This external stress is forcing the displacement followers inwards, so unless the seating of the followers at the zero state is perfect, there will be a stored error movement. At the point where the internal pressure matches the external stress, these imperfections are revealed as a characteristic "signature" for the individual arms. In a simplistic approach, these signatures could be considered as positive indications of the reference pressure. However, in the

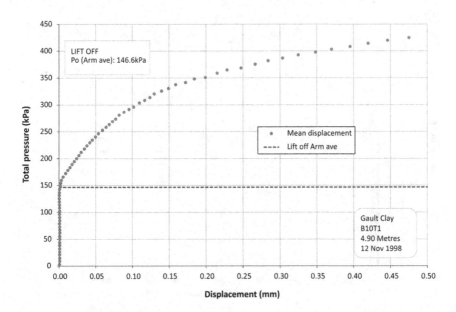

Figure 3.2 An example of lift-off.

ground, it is not possible to have displacements without a consequent change in stress, which will add to, or subtract from, the reference pressure.

As a consequence of these error effects, applying the lift-off analysis means that there can be considerable uncertainty attached to identifying a plausible reference pressure. Conventional practice for coping with this uncertainty is to relax the definition of "lift-off" to mean something more like "significant movement."

Fig. 3.3 is an illustration of the problems involved with identifying lift-off. Here, the individual arms from the SBP test in Fig. 3.2 are plotted.

There are several choices of lift-off stress depending on how rigorous the interpretation of what is implied by the term. In general, the rigorous lift-off stress is that obtained from the first arm to move. The variation in the starting stress distribution indicates defects in the installation process. The probe may not be perfectly vertical, or may not be removing material efficiently enough to avoid raising the local state of stress in material immediately adjacent to the pressuremeter.

It is important to bear in mind the scale. The lift-off information is concentrated into the first 100 micrometres of the expansion or about 0.25% cavity strain. This is less than the strain required to make the material reach the fully plastic condition (yield). Because the movements are well within the sub-yield range of the material, the analyst is justified in attributing significance to the output of the separate arms. In this event, the arithmetic mean of the separate lift-off points can be a more useful parameter than lift-off derived from averaged arm displacement data.

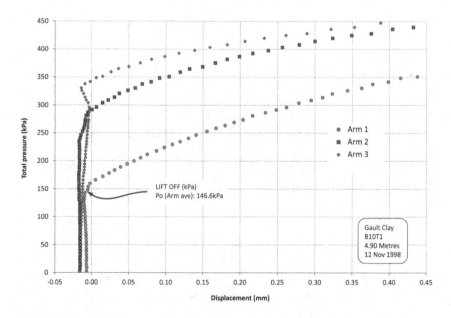

Figure 3.3 Lift-off, all arms shown.

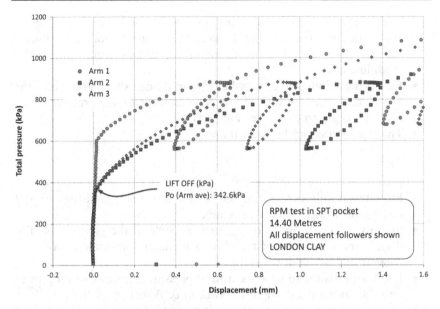

Figure 3.4 Lift-off in a pushed RPM test.

If a strict definition of "lift-off" could reasonably be applied, then no assumptions concerning soil response are required. In the less rigorous application that in practice is followed most of the time, it is important that the analyst identifies the onset of plastic behaviour as a guide to deciding that some conspicuous change of form in the loading curve at a lesser stress is likely to be p_o. Perhaps this should be referred to as p_o by inspection.

Fig. 3.4 is an extreme example. This is an extract from an RPM test made in a pocket formed by SPT tools. The material has been grossly disturbed by the SPT and has probably been taken to near-limit conditions before the pressuremeter is positioned. Nevertheless, there are clear examples of "lift-off" stress. It happens for this test that the best-estimate for p_o is 290 kPa (derived by curve modelling) rather than the lift-off value of 343 kPa but under the circumstances this is acceptable ball-park agreement. It is possible that the influence of the lateral geostatic stress leaves a stress witness in the measured response even under conditions of major disturbance. It is an area that invites further investigation.

3.4 MARSLAND & RANDOLPH (1977) YIELD STRESS ANALYSIS

The Marsland and Randolph analysis for undrained soils relies on being able to identify the onset of plastic behaviour, the yield stress, p_f. The argument is as follows:

- In the vicinity of the *in situ* lateral stress, the soil response is simple elastic and therefore the total pressure/cavity strain plot will be linear. Identify this slope (and incidentally, use it to calculate the initial modulus, G_i).
- Elastic behaviour will cease when the undrained shear strength of the soil is reached in the wall of the cavity, and hence the pressure/strain plot will begin to curve (see Fig. 3.5).
- The yield stress can be written as:

$$p_f = p_o + c_u \qquad\qquad [3.3]$$

- From this it follows that p_o can be deduced by iteration using a plot of displacement against total pressure.
- Initially a guess is made of p_o, such as 50% of the yield stress. The displacement ordinate of the chosen p_o defines an interim origin to convert displacement to strain.
- It is then possible to produce a total pressure:log shear strain plot to find the undrained strength c_u (Gibson & Anderson, 1961).
- The sum of these two parameters is compared with the selected value of p_f. The choice of p_o is then suitably adjusted and the process repeated until a match is found. It is a straightforward matter to carry out this procedure on the computer.

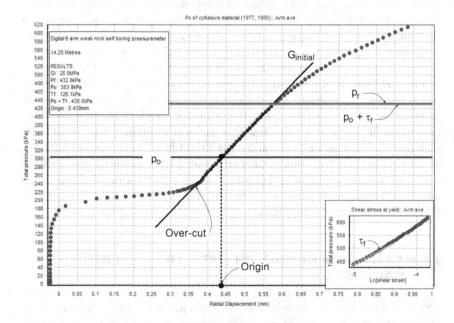

Figure 3.5 An example of the Marsland and Randolph analysis.

The modified method in current use is a response to the difficulty that perfectly plastic deformation is not a realistic enough model for many materials and yield may occur at a different shear stress than the large strain shear strength. Hawkins et al. (1990) suggest that the most appropriate choice is that value of shear stress pertaining at the apparent onset of plasticity, so [3.3] now becomes:

$$p_f = p_0 + \tau_f \tag{3.4}$$

τ_f can be obtained from a total pressure:log shear strain plot by selecting the slope at the pressure and strain corresponding to the choice of p_f. In practice, this substitutes the Palmer (1972) numerical solution for the Gibson and Anderson solution (inset plot in Fig. 3.5).

The analysis is implemented graphically, using rulers to mark break-points on the curve.

It is necessary to be realistic about what the method can do. In the case of the example above (Fig. 3.5), the test is made with a self boring probe arranged to bore fractionally over-size in a stiff clay. The 0.35 mm overcut is very close to the overbore dimension and indicates that the material has not slumped. Under these circumstances, approximating "perfect pre-boring," the analysis gives sensible results because any small relaxation can be supported by the available shear strength. Often this is not the case.

Fig. 3.6 shows a slightly different version of the same analysis applied to a true pre-bored pressuremeter test. The loading was carried out as a series of pressure steps, each step held for one minute. The plot on the left is the

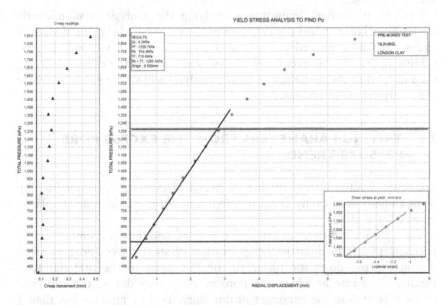

Figure 3.6 An example of the Marsland and Randolph analysis with creep readings.

"creep" movement for each hold. The plot on the right is similar in arrangement as Fig. 3.5 except that the data are those taken at the end of a creep interval so give an especially smooth curve. The creep readings give an additional indication of the yield stress as there is an acceleration in the movement. It is not so easy to detect the cavity reference stress.

The failure condition given in [3.4] could be written:

$$p'_f = p'_o(1 + \sin\phi') \qquad [3.5]$$

where ϕ' is the peak angle of internal friction and stresses are effective. This would make it appropriate for drained tests in purely frictional material such as sand. In place of deciding shear stress from an undrained analysis, a procedure such as Hughes et al. (1977) would need to be applied to find the friction angle.

Alternatively, [3.3] and [3.5] could be combined for a $c'-\phi'$ material:

$$p'_f = p'_0 + p'_0 \sin\phi' + c' \cos\phi' \qquad [3.6]$$

where c' is drained cohesion.

The difficulty is that [3.5] and [3.6] require additional information not directly measured by the pressuremeter test. Equ. [3.4] is easily implemented and applied to all stress paths, some of which imply significant compromise. It will seldom be accurate, but used as a rough guide to the initial stress state, remains useful. However, it is unlikely to be the best tool for the purpose.

For one particular circumstance applying the analysis is significantly misleading. This is when the insertion process has raised the initial state of stress, such as a pushed test but also an under-drilled self-boring pressuremeter test. In this event, the analysis can contribute nothing—forcing such data to fit the assumptions of the analysis will severely over-estimate the cavity reference pressure.

3.5 DERIVING PARAMETERS FROM THE EXCESS PORE PRESSURE TREND

Bolton & Whittle (1999) predict the trend of excess pore water buildup from an undrained cavity expansion in a non-linear elastic/perfectly plastic material (Fig. 3.7).

The significant difference between this trend and that in a simple elastic/perfectly plastic medium is the generation of some excess pore pressure during the pseudo-elastic phase of the test prior to the material fully yielding. The rate at which the pore pressure rises during the sub-yield phase is related to the exponent of non-linearity, β, a number less than 1

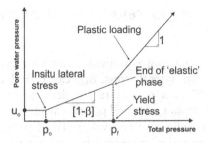

Figure 3.7 The ideal pore pressure response.

unless the response is truly linear elastic (the Bolton and Whittle analysis is described more fully in Chapter 4 and Appendix A).

In both cases, once the material becomes fully plastic, there is a 1 for 1 correspondence between changes in total stress and changes in pore water pressure. In practice, few self-boring tests have the necessary minimal disturbance to show the full theoretical behaviour. Even for tests where the insertion procedure is optimal, interruptions to the loading to take unload/reload cycles tends to disrupt pore water pressure generation.

Fig. 3.8 is a typical example of what is recorded. In the latter stages of the loading when unload/reload cycles are taken, the pore water pressure response levels off. This indicates partial drainage, possibly not in the soil mass, but locally at the borehole wall where gaps in the protective sheath introduce axial drainage paths.

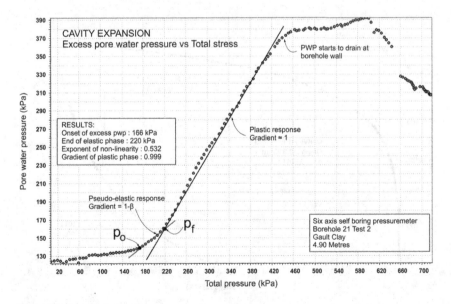

Figure 3.8 An example of pore pressure response.

3.6 DERIVING *IN SITU* LATERAL STRESS BY CURVE MODELLING

The uncertainty in associating a particular value for cavity reference pressure, p_o, with the *in situ* lateral stress, σ_{ho}, can be reduced by curve modelling. Jefferies (1988) is a procedure for deriving in situ lateral stress, stiffness and strength from undrained pressuremeter curves by matching the measured data points with an iteratively selected parameter set. Some rigour is introduced into the procedure by making the set of parameters match the contraction as well as the expansion phases of the SBPM test.

The model used to represent the deformation characteristics of the soil has to be realistic. Jefferies (1988) follows Gibson and Anderson in assuming a simple elastic response until the full shear strength of the material is mobilised and a perfectly plastic response thereafter. Outside of a computer, there is no such soil and the model does not predict the measured field values for stiffness. This is a serious weakness because it is the one property of the soil that pressuremeters provide reliably without major difficulty.

However, the procedure can be used with more representative soil models, and it is customary now to back-analyse undrained tests using a non-linear elastic/perfectly plastic shear stress/shear strain solution. As described in Whittle (1999), this uses as input the already determined values of stiffness and shear strength so the only variable to be decided is the *in situ* lateral stress. Both expansion and contraction phases of the test are fitted (Fig. 3.9). The model is given in more detail in Chapter 4.

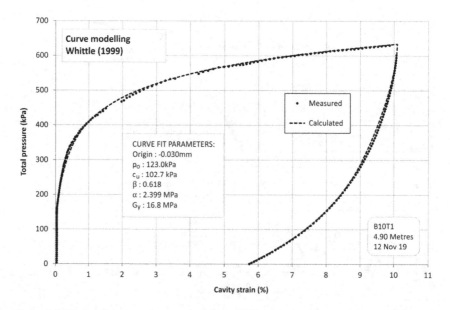

Figure 3.9 Undrained curve fitting example, self-bored test.

There are similar methods for drained tests using a non-linear version of the solution suggested by Carter et al. (1986). These models are outlined in Chapter 5.

For a SBP arranged to drill to size, the values for lateral stress derived using the non-linear undrained model are often <u>lower</u> than those obtained by inspection, and are consistent with a view of the test as slightly under-drilled. This raises the state of stress around the probe. If the probe is configured to drill fractionally oversize, the reverse situation can happen. It is generally easier to interpret an over-drilled test compared to an under-drilled test.

Note that the example is from the same test as the "lift-off" plot given in Fig. 3.2. A slightly higher value for p_o was obtained by that method. This is in line with the model results, which indicate a small negative origin.

Only one value for *in situ* lateral stress is derived using these procedures, as isotropy of soil properties is a fundamental assumption. Because the procedure makes use of all the evidence, it is the preferred method for deriving the *in situ* lateral stress.

3.7 BALANCE PRESSURE CHECK TEST

The balance pressure check (BPC) method is outlined in Hoopes & Hughes (2014), and Fig. 3.10 is an example of a test in clay that includes an implementation of the method. Fig. 3.11 shows the BPC part of the test expanded.

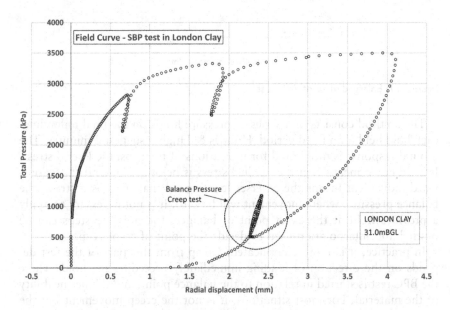

Figure 3.10 SBP test in London Clay that includes a BPC test.

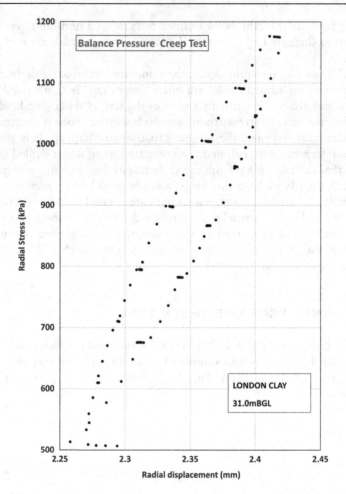

Figure 3.11 Enlarged view of BPC test.

The method consists of a series of pressure holds on the final unloading, each hold held for a fixed period. Hoopes & Hughes suggest 2 minutes. The ground response is monitored for indications of the geostatic lateral stress. Inward creep means that the applied stress is below the lateral stress; outward movement means the reverse. It is argued that there is a stress, the balance pressure, where movement will be zero. It is not necessary to apply a pressure hold at the exact point of balance, the pressure steps can be plotted and interpolation used to decide the point of zero creep.

In practice, it is more complicated. Creep from this part of the test depends on the direction of loading immediately preceding the hold, where the BPC test is started in relation to the balance point, and the permeability of the material. For most situations, it is not the creep movement but the creep rate that becomes zero at the balance point (Fig. 3.12).

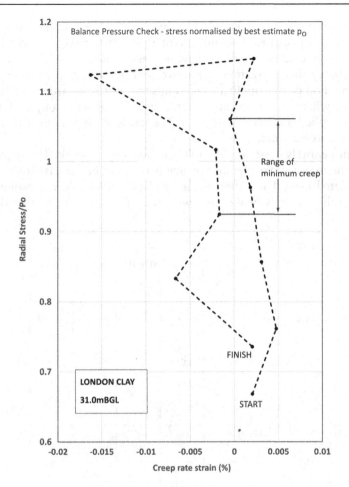

Figure 3.12 BPC interpretation.

The procedure used in this example is to begin the BPC test at approximately 50% of the over-burden stress and continue it until a stress of twice the over-burden has been applied. It is expected that in this material this will cover the plausible range for the *in situ* lateral stress. There are several holds, both on the loading and unloading arm of the BPC test, but held for a relatively short time of 30 seconds. The start and finish of holds give a creep displacement for a particular stress and the difference between adjacent steps is plotted in Fig. 3.12. The stress data have been normalised by the best estimate available for the cavity reference pressure, which in this case was decided by curve modelling.

There is an excellent degree of agreement between the best estimate value and the indications from the creep data.

What is not obvious in Fig. 3.11 is that because the process starts below the vertical stress, there is a null point when this stress is reached. This should not be confused with the lateral stress indicator, which occurs later and probably shows greater creep movement but at a zero creep rate.

The method can be applied to tests in sand, but it is harder to recognise the point where the creep rate is zero. Fig. 3.13 is an example of a BPC analysis applied to a very dense sand. The test was carried out with a small pre-bored pressuremeter.

It is not entirely clear why the BPC method should work. The suggestion is that the material has memory, which is likely to be the case for an over-consolidated clay. This, after all, is the basis of sample reconsolidation. Procedurally, what is being done with the BPC is similar to the Ménard

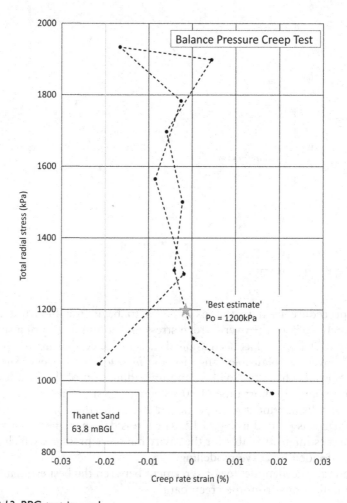

Figure 3.13 BPC test in sand.

methodology of applying pressure increments at the start of the test and waiting for a fixed period. The difference is the degree of cavity unloading that precedes the creep. Following pre-boring, the borehole has been completely unloaded. If the material is soil, then the consequence is irrecoverable disturbance and very little can be determined until the effects of that have been erased.

This requirement is satisfied by a large expansion of the cavity and a subsequent controlled unload down to the start of the BPC procedure. The cavity wall is always supported by the expanded surface of the pressuremeter.

In Fig. 3.10 for example, the BPC cycle has a similar stiffness to the unload/reload cycles carried out on the loading. This is not the case for the initial loading of a pre-bored cavity, where the creep due to moving either side of the geostatic stress potentially is overwhelmed by the magnitude of the creep deformation caused by the process of restoring the cavity to its original dimension.

Nevertheless, despite these cautions, it is possible that creep readings on the initial loading can give a meaningful response. Fig. 3.14 is extracted from the same pre-bored test as that shown in Fig. 3.13.

The magnitude of the creep is much less for the final unloading. However, the initial loading is showing a change of direction in the vicinity of the geostatic stress that would give a zero creep rate. This is not the stress point that would be identified as cavity reference pressure in the standard interpretation of such data.

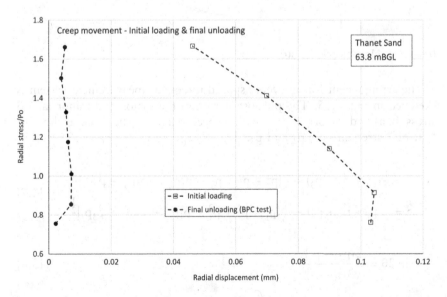

Figure 3.14 Initial loading and final unloading creep compared.

3.8 LATERAL STRESS ANISOTROPY

It is common in rock to see a major and minor lateral stress and in principle there is no reason why the horizontal *in situ* stress in soils should be isotropic. Natural features such as a nearby river or valley may influence the stress field. In material that was once under an ice sheet there may be a bias in one direction due to the path followed by glacial movement.

Dalton & Hawkins (1982) argue that the dispersal of lift-off pressures in a plot such as Fig. 3.3 potentially is due to lateral stress anisotropy. Each reading of "lift-off" stress is associated with an orientation and hence can be reduced to a maximum and minimum with direction using a Mohr's circle calculation.

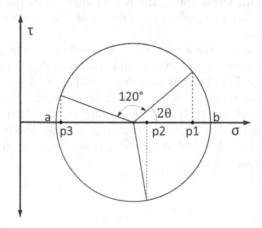

Figure 3.15 Mohr's circle for stress.

The arrangement for an evenly spaced three-axis measurement system is sketched in Fig. 3.15. The purpose is to find the major and minor lateral stress (b, a) and the orientation with respect to axis 1. The three readings of "lift-off" stress are p_1, p_2 and p_3:

$$b = \tfrac{1}{3}(p_1 + p_2 + p_3) + \tfrac{2}{3}\left(p_1^2 + p^2 + p_3^2 - p_1p_2 - p_2p_3 - p_3p_1\right)^{\frac{1}{2}}$$

$$a = \tfrac{1}{3}(p_1 + p_2 + p_3) - \tfrac{2}{3}\left(p_1^2 + p^2 + p_3^2 - p_1p_2 - p_2p_3 - p_3p_1\right)^{\frac{1}{2}}$$

[3.7]

$$\sin 2\theta = \frac{2(p_2 - p_3)}{\sqrt{3}(b - a)}$$

[3.8]

The stress σ at any bearing θ (referred to axis 1) in a horizontal plane is obtained using:

$$\sigma = \frac{(a + b)}{2} + \frac{(b - a)\cos 2\theta}{2}$$ [3.9]

The argument is that a natural anisotropic stress distribution is possible because the shear strength of the material can support small differences.

There is now a greater appreciation that in the presence of small strain stiffness, any disturbance at all will compromise the direct measurement of the geostatic stress. It is unlikely that the experimental data used by Dalton and Hawkins would pass the test of insignificant disturbance. This doesn't prevent the analysis itself from being correct, and in other areas it can be useful.

One such has been the use of a pressuremeter in the horizontal plane, such as in the face of a tunnel boring, where the vertical direction is aligned with one axis.

For tests in rock where a pre-bored pressuremeter is used conventionally, it is also possible to use the Dalton and Hawkins approach to find the ratio σ_H/σ_h and the orientation of σ_H with respect to axis 1 without resolving the magnitude of either stress. Normally, it is the average of all displacement followers that is used in the test interpretation because this is the reading least likely to show movement of the probe axis with respect to the borehole. If one can be confident that the probe is locked in place, then the individual axes can have something to offer, and the obvious place where this can apply is during an unload/reload cycle.

Being rock, the slope of such cycles will be linear, so three values for shear modulus, G, in the horizontal plane can be determined. Equ. [3.7] can then be used to find the ratio between them. The high-pressure dilatometer is fitted with a compass and hence orientation of the major axis can be expressed as a bearing.

This procedure asks a considerable amount about the ability of the equipment to detect these small differences. On the positive side, if several unload/reload cycles are taken, the calculation can be repeated and the results checked for consistency. However "consistency" may also be an instrument artefact. One way of deciding this is to rotate the probe between two closely spaced tests and check that the orientation of the major axis has not also rotated.

A self-boring pressuremeter with six displacement sensors evenly distributed in one plane may be capable of seeing anisotropy in the presence of certain kinds of insertion disturbance. For example, if the axis of drilling is not quite vertical, the probe will follow a curved path with raised stresses on one side and stress relief on the other. Averaging the lift-off readings from opposing pairs of arms potentially cancels out such effects. The averaged three axes[1] give the input for the Dalton and Hawkins analysis, permitting anisotropy parameters to be derived with disturbance effects neutralised.

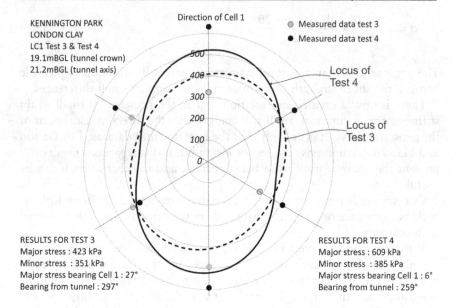

Figure 3.16 Anisotropic stress adjacent to an old tunnel (Gourvenec et al., 2005).

Whether this approach adequately compensates for the stress changes due to disturbance is arguable. It will not apply if disturbance has raised the state of stress everywhere around the instrument.

A situation where anisotropy might be expected is shown in Fig. 3.16. These tests are reported in Gourvenec et al. (2005). The borehole was 1.5 metres from a 4-metre tunnel that has been shield bored about 75 years earlier. The readings of lateral stress at 60° intervals are taken from a six-axis load cell placed by a self-boring technique.

The bearing with respect to the tunnel is suggesting that the major stress is almost normal to the tunnel alignment with a substantial increase in anisotropy at the tunnel axis. Whether such data are robust enough to withstand critical examination is a separate question; the point of the example is to demonstrate the approach.

3.9 A NOTE ABOUT K_O — SUBMERGED MEASUREMENTS

The principle of effective stress is fundamentally important in soil mechanics. It must be treated as the basic axiom, since soil behaviour is governed by it.

Changes in water level below ground (water table changes) result in changes in effective stresses below the water table. Changes in water level above ground (e.g. in lakes, rivers, etc.) do not cause changes in effective stresses in the ground below.

The coefficient of earth pressure at rest can be defined as that state of stress equilibrium where there are no strains in the lateral direction. If the effective vertical stress is unaffected by changes in the water level above ground level then the same must be true of the effective lateral stress; otherwise, the condition of no lateral strain will be violated. Therefore whatever water pressure u is being deducted from the total vertical stress to give effective vertical stress is the same u that must be deducted from the total lateral stress. http://environment.uwe.ac.uk/ geocal/SoilMech

The statement above is a text-book description of the relationship between effective stress and the coefficient of earth pressure at rest, k_o. The ambient water pressure, u_o, is deducted from both the total *in situ* vertical stress, σ_{vo}, and the total *in situ* horizontal stress, σ_{ho}, to give k_o:

$$k_0 = (\sigma_{ho} - u_0)/(\sigma_{vo} - u_0) \qquad [3.10]$$

When applying this to a submerged soil, it is necessary to be clear about what is measured by the pressuremeter to get the appropriate parameters for the calculation.

Fig. 3.17 is the arrangement for a submerged soil, where u_a is the water pressure above sea bed calculated from the water depth z_a and u_b is the water pressure below sea bed calculated from the soil depth, z_b.

The total *in situ* vertical stress is the weight of everything above the element of soil in Fig. 3.17:

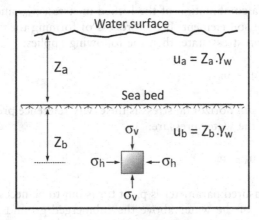

Figure 3.17 Effective stresses, submerged case.

$$\sigma_{vo} = \gamma_s z_b + u_a \qquad\qquad [3.11]$$

where

γ_s is saturated unit weight of the soil
u_a is the water stress above the sea bed $= z_a \gamma_w$
γ_w is the unit weight of water

The effective *in situ* vertical stress is σ_{vo} less the contribution of the water:

$$\sigma'_{v0} = \sigma_{vo} - u_a - u_b \qquad\qquad [3.12]$$

where
u_b is the water stress below the sea bed $= z_b \gamma_w$

Because u_a appears in [3.11] and [3.12], it cancels out in the calculation of effective vertical stress:

$$\sigma'_{v0} = \gamma_s z_b - u_b \qquad\qquad [3.13]$$

Alternatively

$$\sigma'_{v0} = z_b(\gamma_s - \gamma_w) \qquad\qquad [3.14]$$

Equations [3.13] and [3.14] make no mention of the water above the sea bed and satisfy the condition that the *effective* vertical stress is unaffected by changes in the water level above bed level. However, unlike [3.12], they cannot be re-arranged to find the *total* vertical stress when the material is submerged.

A similar argument applies to the *total in situ* horizontal stress, σ_{ho}. The stress determined from the pressuremeter test, referred to as cavity reference pressure, p_o, is a combination of the locked-in stress, σ_{ho}, and the head of water above the soil element. If the process of forming the cavity has not altered the *in situ* stress state, then the following applies:

$$p_0 = \sigma'_{ho} + u_a + u_b \qquad\qquad [3.15]$$

The effective *in situ* horizontal stress is the cavity reference pressure less the contribution of the water pressure:

$$\sigma'_{ho} = p_0 - u_a - u_b \qquad\qquad [3.16]$$

Because the measured parameter is p_o, it turns out to be necessary to know u_a, the pressure of the water above the submerged ground surface. k_o is calculated by combining [3.12] and [3.16], giving:

$$k_o = \frac{P_o - u_a - u_b}{\sigma_{vo} - u_a - u_b} = \frac{\sigma'_{ho}}{\sigma'_{vo}} \qquad [3.17]$$

When calculating the water contribution (u_a and u_b), the unit weight will not always be that of fresh water (≈ 9.81 kN/m^3). Salt or briny water will have a higher unit weight, typically 10.05 kN/m^3.

NOTE

1 If six evenly arranged displacement followers are numbered sequentially in a clockwise direction, then averaging opposing pairs changes the order of the three axes from a clockwise to an anti-clockwise rotation. The right-hand side of [3.8] must be multiplied by –1 to make the orientation calculation correct.

REFERENCES

Baguelin, F., Jezequel, J.F. & Shields, D.H. (1978) *The Pressuremeter and Foundation Engineering*. Transtech Publications, Clausthal, Germany. ISBN 0-87849-019-1

Bolton, M.D. & Whittle, R.W. (1999) A non-linear elastic/perfectly plastic analysis for plane strain undrained expansion tests. *Géotechnique* 49 (1), pp. 133–141.

Carter, J.P., Booker, J.R. & Yeung, S.K. (1986) Cavity expansion in cohesive frictional soils. *Géotechnique* 36 (3), pp. 349–358.

Dalton, J.C.P. & Hawkins, P.G. (1982) Fields of stress- some measurements of the in-situ stress in a meadow in the Cambridgeshire countryside. *Ground Engineering*, May, 15 (4), pp. 15–23.

Gibson, R.E. & Anderson, W.F. (1961) In situ measurement of soil properties with the pressuremeter. *Civil Engineering and Public Works Review* 56 (658), pp. 615–618.

Gourvenec, S.M., Mair, R.J., Bolton, M.D. & Soga, K. (2005) Ground conditions around an old tunnel in London Clay. *Geotechnical Engineering* 158 (1), pp. 25–33.

Hawkins, P.G., Mair, R.J., Mathieson, W.G. & Muir Wood, D. (1990) Pressuremeter measurement of total horizontal stress in stiff clay. Proc. ISP.3 Oxford, ISBN 0 7277 1556 9, pp. 321–330.

Hoopes, O. & Hughes, J.M.O. (2014) In situ lateral stress measurement in glaciolacustrine Seattle clay using the pressuremeter. *Journal of Geotechnical and Geoenvironmental Engineering*, 140 (5),10.1061/(ASCE)GT.1943-5606.0001077

Houlsby, G. & Withers, N.J. (1988) Analysis of the Cone Pressuremeter Test in Clay. *Géotechnique* 38 (4), pp. 573–587.

Hughes, J.M.O. (1973) An instrument for in situ measurement in soft clays, PhD Thesis, University of Cambridge.

Hughes, J.M.O., Wroth, C.P. & Windle, D. (1977) Pressuremeter tests in sands. *Géotechnique*, 27 (4), pp. 455–477.

Jefferies, M.G. (1988) Determination of horizontal geostatic stress in clay with self-bored pressuremeter. *Canadian Geotechnical* 25 (3), pp. 559–573.

Mair, R.J. & Wood, D.M. (1987) Pressuremeter testing. Methods and interpretation. *Construction Industry Research and Information Association Project 335*. Publ. Butterworths, London. ISBN 0-408-02434-8

Marsland, A. & Randolph, M.F. (1977) Comparison of the results from pressuremeter tests and large insitu plate tests in London Clay. *Géotechnique* 27 (2), pp. 217–243.

Palmer, A.C. (1972) Undrained plane-strain expansion of a cylindrical cavity in clay: a simple interpretation of the pressuremeter test. *Géotechnique* 22 (3), pp. 451–457.

Whittle, R.W. (1999) Using non-linear elasticity to obtain the engineering properties of clay. *Ground Engineering* May, 3 (5), pp. 30–34.

Chapter 4

Analyses for Undrained Shear Strength

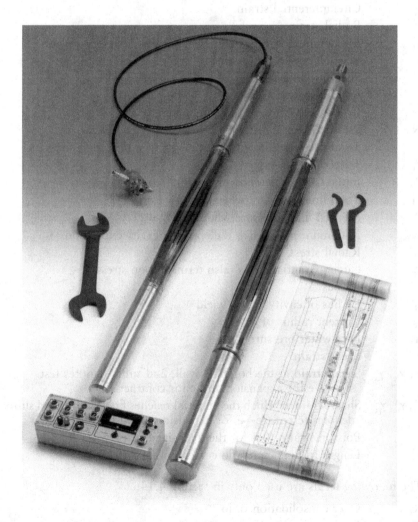

Figure 4.0 N size and H size high-pressure dilatometers.

DOI: 10.1201/9781003200680-4

4.1 NOTATION

A	Current area of cavity
ΔA	Change in current area of cavity
V	Current volume of cavity
ΔV	Change in current volume of cavity
α	Shear stress constant (Bolton & Whittle, 1999)
β	Exponent of non-linearity (Bolton & Whittle, 1999)
ε_a	Axial strain
ε_c	Cavity strain (a specific instance of circumferential strain)
ε_θ	Circumferential strain
ε_r	Radial strain
ε_{max}	Maximum cavity strain at the end of a cavity expansion
E	Elastic (Young's) modulus
G	Shear modulus
I_r	Rigidity index (G/c_u)
k_o	Co-efficient of earth pressure at rest
N_p	Pressuremeter constant, $Ln(I_r) + 1$
P_c	Total pressure measured at the cavity wall (radial stress)
P_f	Total yield stress
P_o	Initial cavity reference stress (total)
P_L	Total limit pressure when $\Delta A/A = 1$
P_{LM}	Ménard limit pressure when $\Delta V/V_o = 1$
P_{max}	Maximum pressure at the end of a cavity expansion
σ_r	Radial stress
σ_θ	Circumferential stress (also termed hoop stress)
r	Radius
r_f	Radius of cavity at first yield
r_c	Current radius of cavity
u	Pore water pressure
γ	Shear strain
γ_c, γ_{ce}, γ_{cc}	Shear strain at the borehole wall, 2nd suffix denotes test stage—e for expansion and c for contraction
γ_γ, $\gamma_{\gamma e}$, $\gamma_{\gamma c}$	Shear strain at which the material reaches first failure, 2nd suffix denotes test stage—e for expansion and c for contraction
υ	Poisson's ratio (0.5 for the undrained case)
N	Length to diameter ratio correction factor

The following terms are used only in section 4.14:

R_p	Over consolidation ratio
P'_γ	Maximum past effective vertical stress

P'_A Current vertical effective stress
M   Frictional coefficient
m   Exponent
r   Spacing ratio
I_P Plasticity index
Λ Plastic volumetric strain ratio
σ'_n Mean effective normal stress
ϕ, ϕ_{ps}, ϕ_{tx} Angle of internal friction, plane strain and triaxial

4.2 INTRODUCTION

Undrained shear strength is not a fundamental material property, it is the measured response to the undrained loading condition. If a cavity expansion and contraction is carried out in an incompressible medium, then strength is independent of the loading stress because any change in the deviatoric stress is carried by pore water pressure. The effective stress following yield is constant and the material deforms without generating volumetric strains.

The remaining strains are straightforward to calculate from the measured displacement of the cavity wall. The expansion and contraction phases of the test independently can be used to determine the current shear stress at any point in the test and at any radius in the soil mass.

If the form of the shear stress/shear strain response is decided in advance, then it is possible to interpret the field data with a closed-form solution. The best known of these for the undrained case is the Gibson & Anderson (1961) solution for a soil response that is linear elastic up to the stress at which the material yields, and perfectly plastic thereafter (Fig. 4.1). This kind of solution is reversible, so that a set of parameters can be used to predict the field data.

However, it is not necessary to make any assumptions about the form of the shear stress curve. In the more rigorous approach, the current slope of the field curve is used to solve the governing differential equation. An example of this is the Palmer (1972) numerical solution for undrained cavity

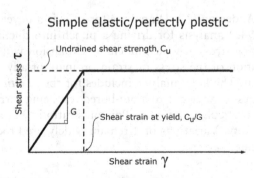

Figure 4.1 Gibson & Anderson 1961.

expansion. Such a solution finds the current mobilised shear stress, but cannot be reversed to reconstruct the experimental data.

The following account of how the undrained test is interpreted starts with simple elastic/perfectly plastic solutions for loading and contraction data. These are then modified to show a non-linear elastic response at sub-yield strains, using a power law. Because the form of the non-linear response is pre-determined, it can be used to solve Palmer, thus incorporating elements of the numerical analysis into closed-form solutions. A hyperbolic alternative to the power law is also considered.

Loading and unloading are brought together in a curve comparison model that despite its simplicity is capable of reproducing the measured data. If the match is poor, then this signifies that either the set of input parameters is not the optimum or the model is unsuitable for the material under test. The parameters can be optimised using an iterative technique that in practice has only one degree of freedom. This constraint means that the model can quantify and correct for the consequences of insertion damage. It remains an engineering decision as to whether the final fit is a sufficiently close match to be considered representative.

This approach is then applied to solutions for the full displacement test and show that it is possible to obtain sensible data for strength and the *in situ* lateral stress despite the total disturbance associated with the pushed insertion technique.

Perfect plasticity is a special case. Lightly over-consolidated soft clays in particular show a strain softening response that is not reproducible by current closed-form solutions but is readily followed by numerical methods. Peak strength will occur at shear strain levels between 1% and 2%. This strain level is unlikely to be preserved by even the best sampling techniques and thus peak strength is difficult to determine except by the cavity expansion test. The lack of comparable data has contributed to a perception that pressuremeters over-estimate shear strength. This issue is examined in some detail.

4.3 CLOSED-FORM SOLUTIONS

The Gibson & Anderson (1961) analysis for soils is a development of the Bishop et al. (1945) analysis for driving a punch into ductile metal. This identifies a limiting stress at which indefinite expansion will occur and allows the calculation of the stress or strain ordinate of any point prior to reaching the limit. The later analysis includes for the *in situ* state of stress and an elastic stress change due to a pre-bored test being carried out. It is a solution for the particular case when Poisson's ratio is 0.5, a pre-requisite for incompressibility. Variations of it remain widely used today.

At the limit state:

$$P_L = P_o + c_u \left(1 + \log_e \left[\frac{E}{2c_u(1 + v)} \right] \right)$$

[4.1]

For a total radial stress, P_c, at the cavity wall that is less than the limit state:

$$P_c = P_o + c_u + c_u \log_e([E/2c_u(1 + v)]\Delta V/V - (1 - \Delta V/V)P_o/c_u) \qquad [4.2]$$

The form in which [4.2] is generally applied is the version due to Windle & Wroth (1977) for a perfectly installed pressuremeter. Young's modulus is replaced by shear modulus, the undrained assumption removing some of the terms in [4.2]:

Windle & Wroth (1977):[1]

$$P_c = P_o + c_u\left[1 + \log_e\left(\frac{G}{c_u}\right) + \log_e\left(\frac{\Delta V}{V}\right)\right] \qquad [4.3]$$

The limit case for indefinite expansion, when $\Delta V/V = 1$, is given by:

$$P_L = P_o + c_u(1 + \log_e[G/c_u]) \qquad [4.4]$$

Conveniently, this can be combined with [4.3] and re-arranged to give the following:

$$P_c = P_L + c_u \log_e\left(\frac{\Delta V}{V}\right) \qquad [4.5]$$

From [4.5], the ultimate gradient of a plot of total pressure against the log of the current cavity shear strain gives the undrained shear strength, and the intercept is the limit pressure (Fig. 4.3).[2]

$\Delta V/V$ is current shear strain measured at the cavity wall, γ_c. The choice of nomenclature depends on the method by which cavity displacement is being determined. Hence:

$$\gamma_c = \Delta A/A = \Delta V/V = 1 - 1/(1 + \varepsilon_c)^2$$

G/c_u is also a shear strain, specifically the inverse of the shear strain, γ_{ce}, at which the material first yields. An alternative way of writing [4.3] using the yield stress is:

$$P_c = P_f + c_u \log_e(\gamma_c/\gamma_{ce}) \qquad [4.6]$$

For a simple elastic material:

$$P_f = P_o + c_u \qquad [4.7]$$

It will be found in subsequent derivations that [4.6] does not change. Only the definition of the yield stress, p_f, and yield shear strain, γ_{ce}, will alter depending on how the sub-yield response of the ground is described.

Following yield, undrained expansion implies that every increment of total pressure, P_c, is matched by an increase in the pore water pressure. Assuming at the start of the test there was an ambient pore water pressure, u_0, in the cavity, the current pore water pressure, u, at the cavity wall is given by:

$$u = u_o + c_u \log_e(\gamma_c/\gamma_{ce}) \tag{4.8}$$

This can be used to calculate the distribution of pore water pressure at all radii up to the radius where the material is on the point of yielding, the elastic/plastic boundary. In practice, [4.8] slightly under-estimates the excess pore water pressure because of soil non-linearity prior to yield. See [4.14].

4.4 NON-LINEAR ELASTIC/PERFECTLY PLASTIC—BOLTON & WHITTLE (1999)

The assumption of linear elasticity up to the point where the full strength of the material is mobilised (yield) is unrealistic. The linear elastic movement of fine grained material such as clays ceases at a shear strain of about 0.001%, but yield occurs at strains of about 1%. Between these two strains, the stiffness in the material reduces, a behaviour misleadingly called non-linear elasticity.

Fig. 4.2 gives the shear stress/shear strain response of a non-linear elastic/ perfectly plastic soil. Both expansion and contraction are included. Bolton

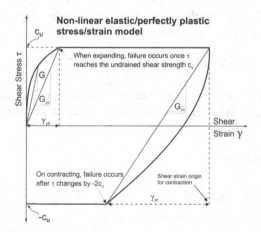

Figure 4.2 Non-linear elastic/perfectly plastic shear stress/strain curve.

& Whittle (1999) assume that the non-linear response of soils can be described by a power law:

$$\tau = \alpha \gamma^\beta \qquad [4.9]$$

where
  τ is shear stress
  α is the shear stress constant
  γ is shear strain
  β is the exponent of non-linearity.

Using parameters that the pressuremeter can measure, it can be shown that the link between radial stress and shear stress is:

$$P_c - P_0 = \frac{\alpha}{\beta}(\gamma_c)^\beta \qquad [4.10]$$

This applies until the shear strength of the material is fully mobilised and the material yields at a radial stress, p_f:

$$P_f = P_0 + \frac{c_u}{\beta} \qquad [4.11]$$

Thereafter, there is a plastic zone confined by the limiting elastic radial stress of p_f. Combining this with the assumption of perfect plasticity leads to the following:

$$P_c = P_o + \frac{c_u}{\beta} + c_u \log_e(\gamma_c/\gamma_{ce}) \qquad [4.12]$$

This is [4.6], with p_f replaced by [4.11]. The shear strain at yield, γ_{ce} is calculated as follows:

$$\gamma_{ce} = \left(\frac{c_u}{\alpha}\right)^{1/\beta} \qquad [4.13]$$

If the material response is assumed to be linear elastic, then $\beta = 1$ and [4.12] reverts to the linear elastic relationship given by [4.3] and [4.6]. Re-defining failure has no effect on the plastic part of the solution so shear strength and limit pressure can be determined from the slope and intercept of a semi-log plot (Fig. 4.3).

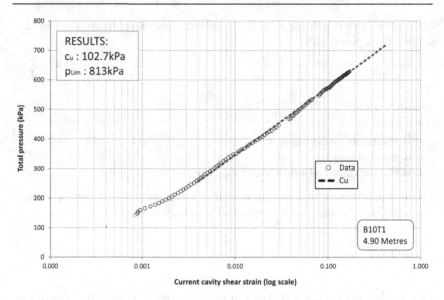

Figure 4.3 Deriving strength and limit pressure from expansion data.

4.5 NON-LINEAR INTERPRETATION OF PORE WATER DEVELOPMENT

One consequence of the non-linear nature of the ground response in un-drained conditions is that some excess pore water pressure development occurs prior to yield. This means that [4.8] becomes:

$$u = u_o + c_u \left[\left(\frac{1 - \beta}{\beta} \right) + \log_e(\gamma_c/\gamma_{ce}) \right] \qquad [4.14]$$

This has implications for the analysis of pressuremeter tests carried out to determine consolidation parameters.

4.6 ANALYSING PRESSUREMETER UNDRAINED CONTRACTION DATA

Jefferies (1988) and independently Houlsby & Withers (1988) apply the linear elastic/perfectly plastic model to the case of an undrained cylindrical cavity contraction. The difference between the two solutions lies only in how strain is defined, the Houlsby and Withers analysis being particularly appropriate for the large strain case.

The expansion phase ends at some value of pressure and cavity strain at the borehole wall, p_{max} and ε_{max}. This is the origin for the contraction event.

During contraction, the end of the elastic phase is reached when the shear stress at the cavity wall has changed by twice the undrained shear strength. Whittle (1999) incorporates non-linearity into this calculation. Hence, the coordinate of radial stress, P_{fc}, and shear strain, γ_{yc}, at which contraction yield occurs, is defined differently, depending on the model adopted:

$$\text{Jefferies (1988):} \quad p_{fc} = p_{mx} - 2c_u \qquad \gamma_{yc} = \gamma_{mx} - \frac{2c_u}{G} \qquad [4.15]$$

$$\text{Whittle (1999):} \quad p_{fc} = p_{mx} - \frac{2c_u}{\beta} \qquad \gamma_{yc} = \gamma_{mx} - \left(\frac{2c_u}{\alpha}\right)^{1/\beta} \qquad [4.16]$$

Note: Contraction shear strain at cavity wall can be derived from cavity strain:
$$\gamma_{cc} = \left(\frac{1 + \varepsilon_{mx}}{1 + \varepsilon_c}\right) - \left(\frac{1 + \varepsilon_c}{1 + \varepsilon_{mx}}\right) \quad [4.17]$$

The Jefferies solution can be written in a generalised form that is applicable to both the linear and non-linear case:

$$p_c = p_{fc} - 2c_u \log_e\left[\gamma_{cc}/\gamma_{yc}\right] \qquad [4.18]$$

An inspection of [4.18] indicates that a plot of log contraction shear strain against total pressure at the cavity wall gives a curve whose ultimate gradient is $-2c_u$. Fig. 4.4 is an example.

As before, if $\beta = 1$, the condition for simple elastic response, all non-linear elastic equations given above revert to published solutions for simple elastic/perfectly plastic material.

Sometimes there is uncertainty in deciding the ultimate slope in both the expansion and contraction examples. For the expansion there can be indications of shear strength changing after the point in loading where unload/reload loops have been taken. It is assumed here that the taking of the cycles has allowed partial drainage, invalidating the primary assumption underpinning the analysis. For the contraction, there is a difficulty in that the slope sometimes increases sharply towards the end. This seems to coincide with the vertical stress becoming the major stress and is a potential issue in soft clay tests with k_o values closer to the normal consolidation state.

A strong argument for using contraction data to discover the shear strength is the certainty of knowing the origin for the contraction event. Unlike the loading state, the strain and stress ordinates of the turn-around are coupled. The uncoupled nature of the loading ordinates means that alterations to the start of the strain calculation for the expansion has a noticeable effect on the derived shear strength. All things being equal, a comparison between loading

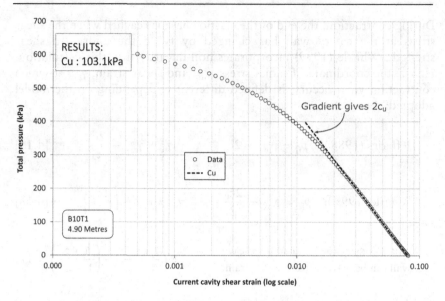

Figure 4.4 Using the contraction curve to derive strength.

and unloading values indicates insertion disturbance and a means of correcting for it. This is a necessary process in implementing the Whittle (1999) undrained model.

4.7 PALMER (1972)

The Palmer analysis[3] is an example of more information being obtained from the pressuremeter test if fewer assumptions are made. The numerical analysis exploits the fact that the field curve is the integrated shear stress response. Taking tangents to the pressure/strain response gives the mobilised shear stress directly, and allows the complete shear stress:shear strain curve to be plotted.

The current shear stress at the cavity wall, τ, is given as follows:

$$\tau = \tfrac{1}{2}\varepsilon_c (1 + \varepsilon_c)(2 + \varepsilon_c)\frac{dp}{d\varepsilon_c} = \frac{dp}{d[\log_e(\Delta A/A)]} \qquad [4.19]$$

If ε_c is small, then [4.19] is approximately:

$$\tau = \varepsilon_c \frac{dp}{d\varepsilon_c} \qquad [4.20]$$

In the absence of a computer[4] this partial differential can be solved using the simple graphical construction suggested by Baugelin et al. (1972) known as the sub-tangent analysis (Fig. 4.5). It was the standard method for deriving shear strength prior to the availability of inexpensive computers.

Using software to take tangents can be troublesome because the differentiation process highlights irregular steps in the experimental data (Fig. 4.6).

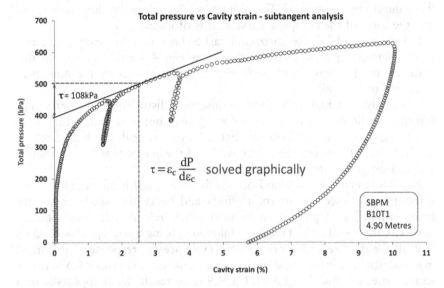

Figure 4.5 Sub-tangent analysis.

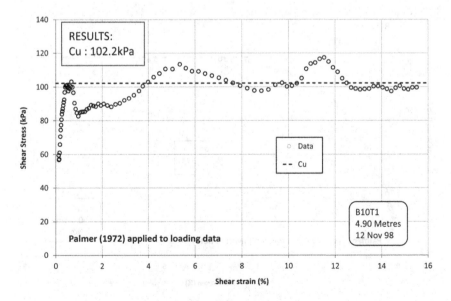

Figure 4.6 Palmer (1972) applied to cavity expansion.

This is frustrating because the stress/strain response must be a smooth curve. The field data can be made regular with an equivalent mathematical expression prior to applying the solution (i.e. the Bolton and Whittle analysis of unload/reload cycles) but in general this is unwise—there are subtle changes of gradient that curve fitting tends to mask.

If there are clear indications of peak and residual shear strength then these should be highlighted. The plot gives a "map" of the shear stress, and it is the form of the complete curve that is of interest.

In Figs. 4.6 and 4.7, the horizontal dashed lines are taken from the results of the perfectly plastic analyses shown in Fig. 4.3 and Fig. 4.4. In this material, the assumption of perfect plasticity is shown by the numerical approach to be justified.

The analysis is highly sensitive to insertion disturbance—in particular, insufficient allowance for stress relief will give a contrived peak in the stress/strain response. It is also possible that an apparent peak is a rate effect, as all the material between the cavity wall and the elastic-plastic transition is seeing a different rate of expansion.

Fig. 4.8 is taken from Wood & Wroth (1977) and is an example of insufficient allowance for insertion effects and hence the wrong choice for strain origin. The slope of the attempted unload/reload cycle does not compare with the initial part of the final unloading, being too steep. This suggests that all the data recorded prior to the cycle are the response of disturbed material. Right shifting the strain origin a small amount (about 0.5% cavity strain), loses the misleading data. Fig. 4.9 is the result. What appears to be a

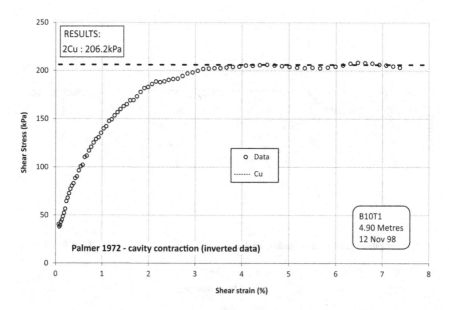

Figure 4.7 Palmer (1972) applied to cavity contraction.

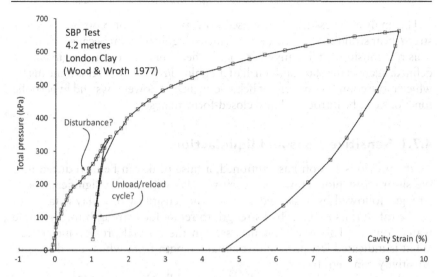

Figure 4.8 Self-bored test with potential disturbance.

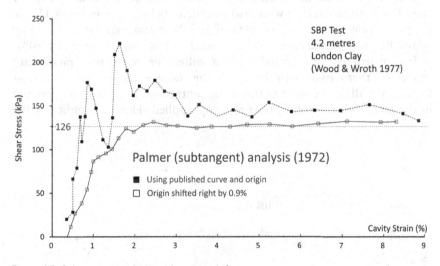

Figure 4.9 Subtangent analysis with origin shift.

peak in the original shear stress response disappears following the origin adjustment and the shear stress curve becomes regular around an average value of 126 kPa. The original published trend is plotting higher than this at all points and is one of several reasons why the perception has been created that pressuremeters over-estimate shear strength.

Because the origin for the contraction event is more or less known and for procedural reasons (there tends to be fewer interruptions to the unloading phase), applying the subtangent analysis to the unloading of the cavity usually gives the smoother result.

The analysis is less frequently used now as a means of plotting the shear stress/shear strain curve. However, it remains highly relevant because of [4.20]. This relationship applies anywhere on the field curve where an origin can be defined, such as the start and finish of an unload/reload cycle. It is a means by which more complex patterns of behaviour such as power laws and hyperbolic functions can be introduced into closed-form analyses.

4.7.1 Sensitive clays and liquefaction

As the previous section has cautioned, a false peak can be introduced into the shear stress plot. However, sensitive clays will show a significant peak strength followed by a marked decline. The sensitivity of a clay is taken to be the ratio of its undisturbed strength to remoulded strength for the same water content. This is not what is seen in the normal pressuremeter test. What is indicated is a peak and residual strength from which the degree of sensitivity can be inferred.

Fig. 4.10 is an early self-bored test reported by Hughes (1973) and is one of a series carried out at the Ellingsrud test site near Oslo, Norway, in quick clay. The contraction data were also recorded. As indicated in Fig. 4.11, the expansion phase has a marked peak of 52 kPa at approximately 1% cavity strain. By the time the cavity has expanded to 5%, this has fallen by 60%.

The instrument was fitted with a difference measuring piezometer transducer that moved with the membrane and was always pressed against the cavity wall. The output of this is the subtraction of pore water pressure in the soil from the total pressure being applied. Hence, it indicates the

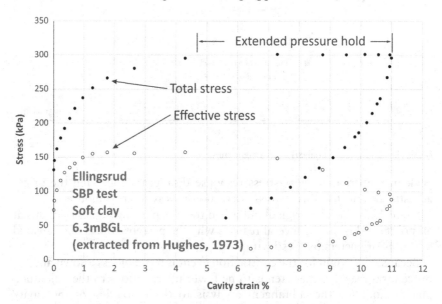

Figure 4.10 Self-bored test in sensitive clay.

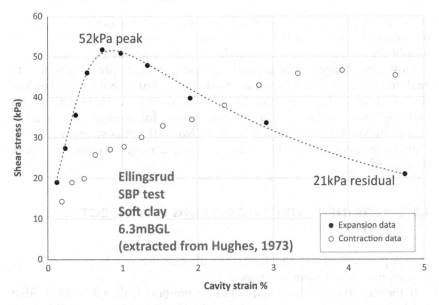

Figure 4.11 Shear stress, expansion and contraction.

effective stress directly. The data show constant effective stress between 1% up to about 5% cavity strain. At this point, the test becomes an extended pressure hold.

If the effective stress trend is correct, then what follows is pore water pressure *increasing* during this holding phase with a consequent loss in effective stress. Whilst the applied total stress remains constant, the material is showing liquefaction behaviour.

When the load is removed and the cavity allowed to contract, this process stops. No such behaviour is evident on the unloading (these data are rotated and mirrored to plot on the same axes as the loading). The ultimate contraction derived shear strength is 44 kPa. If liquefaction has taken place, then the voids ratio of the immediately surrounding ground has been altered and the contraction phase is being conducted in material with different and potentially "improved" properties. Strength is largely restored.

Liquefaction is more common in granular materials that dilate or compress when sheared. Refer to chapter 5.

4.7.2 Appropriate use of subtangent analysis

For most of this chapter, it is assumed that perfect plasticity is a reasonable description of the ground response to undrained shearing. If the soil is strain hardening or more usually softening, as in Fig. 4.11, then the closed-form solutions that assume perfect plasticity are not valid and if adopted, will mislead.

The pressuremeter test is well-behaved, so that even when the strength of the material is reducing it is still necessary to increase the pressure in the membrane to keep the cavity expanding. This is a desirable characteristic—it allows the change of strength to be followed but the only sensible way to analyse such data is to differentiate the field curve at all points and produce the full shear stress/shear strain plot. Trying to force the field data to fit a perfectly plastic solution will result in values for strength that are too great. The initial part of the test is controlled by the peak strength and that information persists. It will be inappropriately averaged if perfect plasticity is assumed.

4.8 UNDRAINED CURVE MODELLING OF PERFECTLY PLASTIC DATA

Perfect plasticity is a special case, the only one that can be completely expressed in a closed-form arrangement.

If the material has been self-bored or pre-bored, then it is likely that producing a model curve to compare with the measured data will be possible. Such a model can identify the extent of the disturbance effect and allow the parameter set to be optimised.

If the material has been pushed, then loading strain data will be indeterminate because it has been completely disturbed. However, if at the end of the test the material is at limit pressure, the construction suggested by Houlsby & Withers (1988), with some modification, can be applied. This utilises the contraction data only but nevertheless is capable of recovering a complete parameter set for strength, stiffness and potentially the *in situ* lateral stress, σ_{ho}.

4.8.1 Modelling the undrained test—Whittle (1999)

Jefferies (1988) is a solution for the full undrained self-bored pressuremeter curve that permits a model curve to be derived from a set of parameters for strength, stiffness and *in situ* stress. It assumes a linear elastic/perfectly plastic stress/strain response. Because linear elasticity up to the point of yield does not give a sufficiently accurate measure of the yield coordinate in loading and unloading the model is insufficiently constrained. In particular, it cannot relate the measured shear modulus to the stiffness value required by the model.

Whittle (1999) is an adaptation of the model that allows a non-linear response up to yield. It is a solution with only one degree of freedom, which is σ_{ho}.

The equations for the non-linear model have already been given but are presented here again for clarity. There are four stages and the calculations start from the cavity reference state when all strains are zero and the total pressure at the cavity wall is p_o:

Stage	Total stress	Shear strain	Stop condition
"Elastic" expansion:	$P_c = P_0 + \left(\dfrac{\alpha}{\beta}\right)(\gamma_c)^\beta$	$\gamma_c = \left[\dfrac{(P_c - P_0)\beta}{\alpha}\right]^{1/\beta}$	$P_c - P_0 = \dfrac{c_u}{\beta}$
Plastic expansion:	$P_c = P_0 + \dfrac{c_u}{\beta} + c_u \log_e(\gamma_c/\gamma_{ce})$		$P_c = P_{mx}$ $\gamma_c = \gamma_{mx}$
"Elastic" contraction:	$P_c = P_{mx} - \left(\dfrac{\alpha}{\beta}\right)(\gamma_{cc})^\beta$	$\gamma_{cc} = \left[\dfrac{(P_{mx} - P_c)\beta}{\alpha}\right]^{1/\beta}$	$P_c = P_{mx} - \dfrac{2c_u}{\beta}$
Plastic contraction:	$P_c = P_{mx} - \dfrac{2c_u}{\beta} - 2c_u \log_e\left[\gamma_{cc}/\gamma_{yc}\right]$		$P_c = 0$ or $\gamma_{cc} = 0$

There is a subtlety in how the model is implemented that is not obvious in the published text. The non-linear elastic exponent, β, is taken from an unload/reload cycle (usually the last carried out), but the shear stress constant α is calculated. It could be taken from an unload/reload cycle but is derived by rearranging [4.13] and so is linked to the choice of c_u and p_L, forcing the model to use the assumption of perfect plasticity. One of the checks on the quality of the fit is how closely the calculated α agrees with the measured α from the chosen cycle. Fig. 4.12 is an example of the procedure in operation applied to a self-bored pressuremeter test in Gault clay. Apart from stripping out the unload/reload data, this is the full curve.

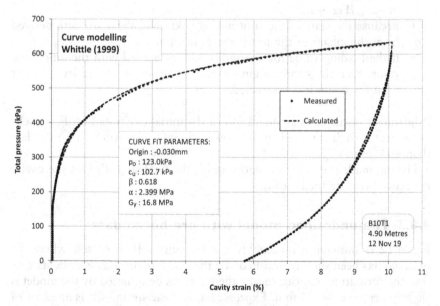

Figure 4.12 Undrained curve modelling—Whittle (1999).

To use the model, the following need to have occurred:

- An initial cavity reference coordinate (the origin) be defined for total pressure and the commencement of the expansion. This coordinate may subsequently change in order to find the best agreement between measured and calculated data.
- At least one unload/reload cycle analysed using the method of Bolton & Whittle (1999) to give non-linear elastic parameters.
- Undrained shear strength and limit pressure decided from the expansion data using the Gibson & Anderson (1961) procedure.
- Undrained shear strength derived from the final contraction derived using the Jefferies (1988) solution.
- Ideally, but not critically, the coordinate of maximum pressure and displacement explicitly identified for when the cavity contraction commences. If not decided by the user, the program defaults to the maximum displacement and maximum pressure reached.

4.8.2 Applying the solution

There are only two steps to implement, based on the following assumptions:

- It is an assumption that strength is the same whether loading or unloading, so yield in contraction occurs after a shear stress change of $-2c_u$ (Fig. 4.2).
- Because the start of the contraction is known, shear strength derived from this phase of the test is the best estimate.
- If the loading value for strength does not accord with the unloading value, then the strain origin for the expansion is moved left or right until the condition is satisfied.

Adjustment means a small alteration of the initial cavity radius. For a self-bored test with disturbance in the sub-yield range, this will be an offset less than ±0.2 mm.

Having made the shear strengths agree, the final step is to set the cavity reference pressure for best fit.

4.8.3 The undrained model with pre-bored data

Fig. 4.12 is showing a reasonably good quality self-bored test where the material has been slightly "pushed" by the pressuremeter as it made its way into the formation. Consequently, the origin as determined by the model is slightly negative, –0.03 mm. Expressed as a shear strain, this is an error of only 0.14%, is easy to allow for and the calculated and measured response show a high degree of correlation.

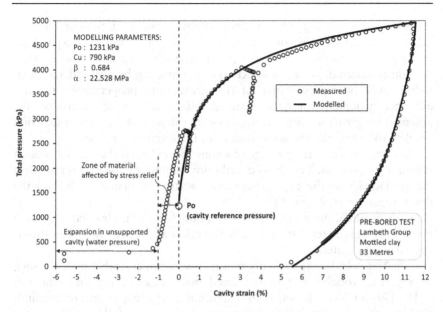

Figure 4.13 Undrained curve modelling a pre-bored test.

The model can also be applied to data where the pocket has been pre-bored. After the removal of the boring tool, the material at the cavity wall will be unloaded elastically and in the case of a soil, plastically.

If it is accepted that the initial part of the test are data that will be complex to match, the undrained model can still be applied albeit with a large initial offset. A greater reliance is placed on the contraction phase of the test. See Fig. 4.13. This is an annotated example of a pre-bored test in a very stiff and strong clay. Although following the removal of the boring tool the cavity wall has contracted plastically, it has not slumped and the modelled data are persuasive.

In weaker material, the initial damage will be more severe and the required strain offset greater.

4.9 HOULSBY & WITHERS (1988)

For the circumstance where the pressuremeter is pushed directly into virgin ground, the models described above are of limited use.

The primary assumption conditioning the analysis is that the installation of the pressuremeter results in the soil experiencing a repeatable amount of disturbance. This be modelled as a cavity expansion from zero initial radius to something approaching the limit condition of the material. The geometry of the hole formation tooling (typically a cone or SPT) is such that some relaxation from this condition takes place prior to the pressuremeter arriving at the same level in the ground. The pressuremeter test needs to reestablish the maximum stress state reached by the hole forming tool and then expand the cavity past this so that all affected soil is experiencing a greater strain than previously suffered. Now the direction of loading can be reversed and what follows will be representative of the stress-strain properties of the undisturbed ground. Only data obtained following stress reversal have the potential for giving realistic parameters—the purpose of the initial expansion is solely to over-write the stress history of the insertion process.

The comparison of the passage of a cone to a cylindrical cavity expansion is an approximation. The stress distribution behind the cone tip is complex, but provided that the pressuremeter is located 10 diameters behind the point, seems to be justified (Teh, 1987).

Houlsby & Withers (1988) is a solution for the undrained case of a pushed pressuremeter test. It specifically takes account of the large strains caused to the material.

If the test is conducted rapidly in a relatively impermeable material such as clay, it is probable that an undrained analysis can be applied to the test.

The Gibson & Anderson (1961) solution for the expansion of a cylindrical cavity from zero initial radius can be written as follows. This is the same as [4.4], with G/c_u replaced by the rigidity index, I_r:

$$p_L = p_o + c_u(1 + \log_e[I_r]) \qquad [4.21]$$

Although the pushed test can establish the limiting pressure, it is not possible to deduce the constituent stress and stiffness elements from the expansion phase of the test. In order to derive such parameters, Houlsby & Withers (1988) extend the undrained analysis to include the contraction of a cylindrical cavity from this limiting pressure. For the contraction, the maximum displacement and pressure at the end of the loading become a new origin.

There are two parts to the analysis—the first is an elastic unloading which ceases when the stress difference exceeds twice the undrained shear strength:

$$\sigma_r - \sigma_\theta = 2c_u \qquad [4.22]$$

where

σ_r is radial stress (the measured pressure applied by the pressuremeter to the cavity wall)

σ_θ is circumferential stress

The strain to the onset of reverse plasticity is:

$$\varepsilon_{max} - (1/I_r) \qquad [4.23]$$

ε_{max} is the maximum cavity strain reached at the end of the expansion part of the test.

During the plastic unloading, the pressure and strain at the borehole wall, p_c and ε_c, are related by the following:

$$p_c = p_L - 2c_u\{1 + \log_e[\sinh(\varepsilon_{max} - \varepsilon_c)] + \log_e[\sinh(I_r)]\} \qquad [4.24]$$

Houlsby and Withers use the hyperbolic trigonometry function sinh because it gives an exact result when its argument is the natural log of a rational number. It is approximately 1 for the strain range being considered, so [4.24] is adequately represented by:

$$p_c = p_L - 2c_u\{1 + \log_e(\varepsilon_{max} - \varepsilon_c) + \log_e(I_r)\} \qquad [4.25]$$

At the end of loading, p_c will equal p_L. It follows that:

$$2c_u[1 + \log_e(I_r)] + 2c_u\log_e(\varepsilon_{max} - \varepsilon_c) = 0$$

hence: $\qquad \log_e(\varepsilon_{max} - \varepsilon_c) = 1 + \log_e(I_r) \qquad [4.26]$

If pressure p_c is plotted against $\log_e(\varepsilon_{max} - \varepsilon_c)$, a construction is obtained from which all stress parameters can be deduced. The slope of the straight line portion of the unloading curve will be approximately $-2c_u$. The intercept of this slope with the pressure at the start of unloading will give $1 + \log_e(I_r)$. Fig. 4.14 gives an example of this graphical construction

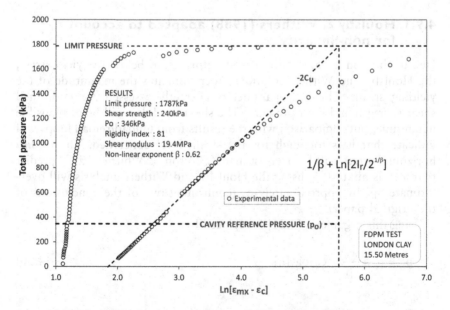

Figure 4.14 An example of Houlsby & Withers (1988), with modifications.

(but with modifications). Knowing I_r allows the shear modulus, G, to be determined, being the product of $c_u I_r$. This shear modulus must be a minimum value because the strains concerned are the maximum elastic strains that the material can support before yielding.

For the purposes of obtaining the undrained shear strength, the construction works, even if the maximum pressure reached at the end of the test is not the limit pressure of the material. However, in order to obtain a value for σ_{ho}, the limit pressure must be measured as the following argument demonstrates:

When $\log_e(\varepsilon_{max} - \varepsilon_c) = 0$ $\qquad p_c = p_L - 2c_u[1 + \log_e(I_r)]$ [4.27]

From [4.21]: $\qquad p_L = p_o + c_u(1 + \log_e[I_r])$

and substituting for the limit pressure in[4. 27] gives: $\qquad p_c = p_o - c_u(1 + \log_e[I_r])$ [4.28]

Adding [4.21] and [4.28]: $\qquad p_c + p_L = 2p_0 = 2\sigma_{ho}$ [4.29]

Hence, the intercept of the semi-log plot at zero strain gives a pressure from which the *in situ* horizontal stress can be determined. In practice, there is no need to draw this; if the limit pressure and rigidity index are known, then σ_{ho} can be calculated from [4.21].

4.9.1 Houlsby & Withers (1988) adapted to account for non-linearity

Due to the non-linear nature of soil deformation below the yield stress, the Houlsby and Withers solution under-estimates the magnitude of the yielding strain. The only parameters correctly assessed are undrained shear strength and limit pressure. The shear modulus will be of plausible magnitude, but comparison with the results from unload/reload loops will indicate that it is too high for the strain level concerned. The *in situ* horizontal stress will be over-estimated. Bolton & Whittle (1999) predict that as β is about 0.5, using the Houlsby and Withers analysis will over-estimate σ_{ho} by approximately c_u (confirming one of the conclusions of the original paper).

Using this argument, [4.21] is replaced by:

$$p_L = p_o + \frac{c_u}{\beta} + c_u \log_e[I_r]$$ [4.30]

The rigidity index, I_r, is the inverse of the initial yield strain and now is defined as follows:

Non-linear rigidity index I_r:
$$I_r = \frac{G_y}{c_u} = \left(\frac{c_u}{\alpha}\right)^{-\left(\frac{1}{\beta}\right)}$$
[4.31]

Similarly, [4.27] is replaced by:
$$p_c = p_L - 2c_u\left[\frac{1}{\beta} + \log_e(I_r/2^\beta)\right]$$
[4.32]

Equ. [4.32] uses the same first yield definition of I_r given in [4.31]. The results indicate that these modifications give plausible values for the cavity reference pressure in materials where the final slope of the unloading is unambiguous, which tends to imply over-consolidated clays.

4.10 ALTERNATIVES TO THE POWER LAW

The concept of redefining the yield criteria to better represent the non-linear response of soils prior to yield is not restricted to a power law. It happens that the power parameters are convenient to manipulate and give an excellent fit to field data, but other researchers have given solutions where a simple hyperbolic function has been used instead.

One such solution is given by Ferreira and Robertson (1992). Their solution uses the following as a description of the pseudo-elastic response (with slightly altered nomenclature):

Loading:

$$\tau = \frac{\gamma}{\dfrac{1}{G_{mx}} + \dfrac{\gamma}{c_u}}$$
[4.33]

Contraction:

$$\tau = \frac{\gamma}{\dfrac{1}{G_{mx}} - \dfrac{\gamma}{2c_u}}$$
[4.34]

where
  G_{mx} is the linear elastic shear modulus at very small strain
  c_u is the undrained shear strength.

Equations [4.33] and [4.34] are treated as constitutive relationships. Using Palmer (1972), it is possible to obtain the connection to the radial stress and build a curve comparison procedure. The authors suggest values for G_{mx} be identified iteratively.

Cao et al. (2002) use the same hyperbolic function (and compare the results directly with the Bolton and Whittle solution). This applies [4.6] with p_f defined as follows:

$$p_f = p_0 + c_u \log_e\left(\frac{G_{mx}}{G_y}\right) \tag{4.35}$$

Note that:

$$\frac{1}{G_y} = \frac{\gamma_{ye}}{c_u}$$

Assuming that the power law and hyperbolic approach give the same result for the yield stress, p_f, it is straightforward to show the connection between G_{mx} and G_y is:

$$G_{mx} = G_y . \exp\left(\frac{1}{\beta}\right) \tag{4.36}$$

The easiest way to apply the hyperbolic functions is to use power law parameters to obtain estimates for the yielding modulus G_y and [4.36] to obtain a value for the elastic modulus, G_{mx}. Fig. 4.15 is an example of

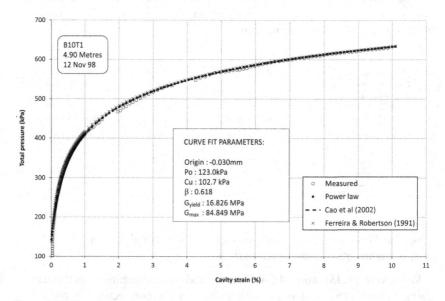

Figure 4.15 Hyperbolic alternative to power law.

several of these curve-matching routines applied to the same experimental data and it is difficult to distinguish between them. Equ. [4.36] seems to be reasonably successful at providing conservative values for G_{mx}.

It is likely that hyperbolic versions of curve matching will prove to be easier to relate to soil parameters derived from laboratory testing.

4.11 STRENGTH FROM LIMIT PRESSURE

Much of the historic pressuremeter strength data found in the literature has been derived using the limit pressure method. Equation [4.4] can be rewritten as:

$$c_u = \frac{P_L - P_0}{N_p} \qquad [4.37]$$

Marsland and Randolph (1977) refer to N_p as the "pressuremeter constant." For a linear elastic material, it is given by $\ln(I_r) + 1$ and a typical value is 6.2.

Wroth (1978) is dismissive of this approach, arguing that the gradient of the semi-log plot (Fig. 4.3) gives the shear strength directly without the need for assessing N_p. However, this judgement is made in the context of self-bored tests with low insertion disturbance and accurate measurement of displacements. For devices that push or stress relieve the material, the easier option is identifying significant pressures. The assumption of a constant rigidity index may then assist in making the best of imperfect data. It depends where lies the greater uncertainty.

Mair and Wood (1987) suggest using unload/reload data to derive an estimate of shear modulus, G, which in turn can be used to find the rigidity index. This introduces an inappropriate level of sophistication into what is an approximate calculation. If the process has any merit, then N_p should be fixed and thus a means of comparing data from different locations.

Powell (1990) presents data from extensive testing in Gault Clay in and around Cambridge interpreted in both ways. One sequence of tests (Fig. 4.16) is from the former test site at High Cross used by the Cambridge University Engineering Department. As Powell points out, for self-bored tests it makes little difference to the regularity of the data how strength is decided. Note the scatter in the 38 mm sample results—the Gault Clay is notorious for showing low strength when tested in the laboratory.[5] The value of N_p is not supplied but inspection of other data in the reference suggests that it is ≈ 5.3. If this method is used, the N_p value should always be given with the derived shear strength.

The limit pressure approach tends to give a conservative trend compared to strength obtained from the gradient. A perfectly plastic fully undrained response throughout the cavity expansion is rarely achieved. Even if P_o and N_p are correct, as fissures and other discontinuities come into play, the

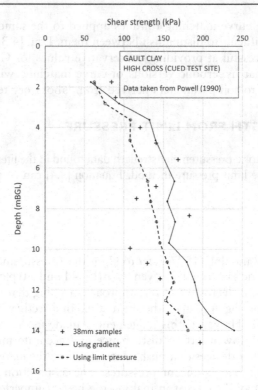

Figure 4.16 The limit pressure method (Powell 1990).

extrapolated limit pressure will be less than the maximum that would be inferred from knowing the stress and strain at which the material first yields.

4.12 PRESSUREMETERS OVER-ESTIMATE SHEAR STRENGTH?

It has been argued that a finite length pressuremeter must over-estimate the material strength. At the ends of the expanding section, an element of soil experiences a small axial component of deformation. It follows that the work undergone when expanding the cavity is under-represented if quantified using only radial displacements. The effect is less if displacements are measured at the centre of the membrane but is still potentially significant.

Houlsby & Carter (1993) carried out tests in a linear elastic/perfectly plastic medium using finite element modelling (FEM) and demonstrated significant L/D (length to diameter) effects. Many researchers have replicated the experiments (for example Shuttle & Jefferies, 1995) and the general conclusion is that field values for shear strength obtained with the pressuremeter should be reduced by a factor somewhere between 25–50%, depending on the rigidity index of the material.

If the predictions of the finite element studies are valid, then the strain dependent nature of the finite geometry effect will produce stress/strain curves from field tests with a strain hardening response. The degree of this will depend on a number of factors, but for a given pressuremeter the significant variable will be the rigidity index.

The Shuttle and Jefferies research offers a convenient way to access the finite geometry results. At any point in the test they argue:

$$c_u^{true} = Nc_u^{meas} \qquad\qquad [4.38]$$

N is a strain dependent variable and is calculated from:

$$N = 1.241 - 0.05\left[\ln\left(I_r\varepsilon_c\right)\right] \qquad\qquad [4.39]$$

The cavity strain, ε_c, is given in percent.[6]

It is possible to turn these two equations into a predicted ground response by recalculating the pressures to find the trend that, when corrected, would give the infinite length case. Fig. 4.17 shows the result for a material with a rigidity index of 200.

According to this figure, there is a clear strain hardening trend for the finite length pressuremeter. However, there are no such data in the testing record. As Fig. 4.18 indicates, the typical response in an over-consolidated clay is close to perfectly plastic. These data are from a series of tests in London Clay where the shear stress/shear strain curve has been normal-ising using the best estimate of the undrained shear strength. The data

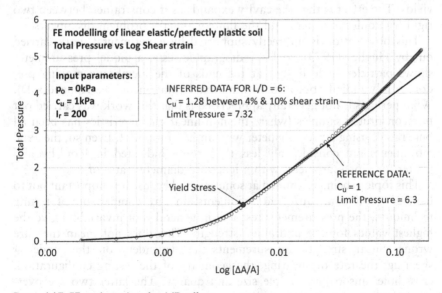

Figure 4.17 FE analysis data for L/D effects.

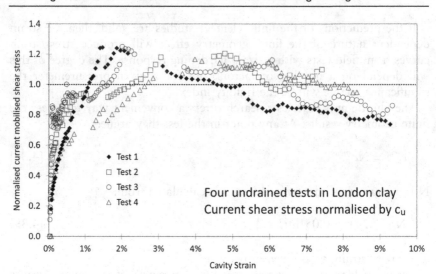

Figure 4.18 Normalised Palmer analysis of four tests.

show some numerical instability, but the general picture is clear. If the FE results were correct for real material, this measured response should not be possible.

The most likely explanation for why the predicted response does not accord with reality is the non-linear stiffness of soils. The stiffness acting at the ends of the expanding membrane where movements are initially small is greater than at the membrane centre where the material first yields. The effect is that the cavity expands as if constrained between two rigid frictionless surfaces and the finite geometry problem disappears.[7]

This observation is indirectly supported by experimental work carried out by Hughes (1973). In his radiographic tests, a model pressuremeter shows particles of lead shot at the ends of the membrane moving predominantly radially but also with an axial component (see Appendix D). What perhaps is not generally appreciated is that this work was carried out in reconstituted samples (where the non-linear characteristics has been all but erased) using a pressuremeter with an L/D ratio of 2. Even so, the axial movements are considerably less than those indicated in Houlsby and Carter for a pressuremeter with a length to diameter ratio of 6.

This topic is being examined at some length because it is important not to damage good data with bad arguments. In any comparison of testing techniques, the pressuremeter test will in general (not invariably) give the highest values for the undrained strength. This does not mean they are wrong. Shear strength measurements are dependent on the mode of shearing, the rate of shearing, the geometry of the testing configuration, cross-hole anisotropy, sample size and quality. The latter two are overwhelmingly the greatest variables.

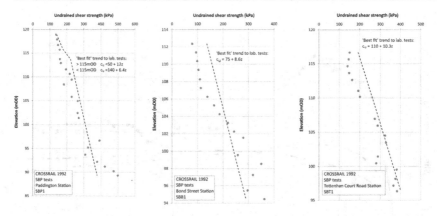

Figure 4.19 Cu profile (a).   *Figure 4.20* Cu profile (b).   *Figure 4.21* Cu profile (c).

Despite these reservations, laboratory and *in situ* tests are not always as different as has been suggested. Fig. 4.19 is taken from testing carried out in 1992 for the London Crossrail project. Extensive self-bored pressuremeter testing was carried out and the results for shear strength are presented together with the best-fit trends from the triaxial testing that was part of the same investigation. Two additional locations are shown in Figs. 4.20 and 4.21. It is difficult to look at these and conclude that the pressuremeter is over-estimating strength.

The data in these figures are from over-consolidated London Clay, where k_o is at least 1. The greatest disparity between pressuremeter strength and other data sources seems to be from tests where k_o is less than 1, tending to the normally consolidated state. It is harder then to find agreement between the different types of testing.

The data from the quick clay in Fig. 4.11 have been compared to the results from vane tests carried out at the same test site and Hughes (1973) comes to the following conclusions:

a. The vane tests give results slightly lower than the *residual* strength data from the pressuremeter tests.
b. The disparity is greatest when the peak to residual ratio is highest.
c. When compared to an induced failure from a trial embankment, the vane values are too low by a factor of at least 2.
d. Only the pressuremeter peak strength values come close to the trial results.

Fig. 4.22 shows data from the former U.K. soft clay test site at Bothkennar. Results from this site have been extensively reported (see, for example, Nash et al., 1992). Self-boring testing was also carried out at the site as late

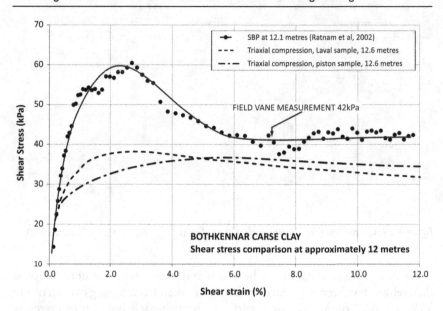

Figure 4.22 Soft clay strength comparison.

as 2000 (Ratnam et al., 2002). Data from the original reporting has been combined with the SBP test results for one depth, approximately 12 mBGL. The triaxial testing trends are from samples consolidated to the assumed k_o conditions.[8]

The peak strength is obtained at shear strains well below the level that an extracted sample will undergo prior to being tested in the laboratory. Potentially, a laboratory test will give a value closer to the residual strength, but likely to be below the shear stress determined *in situ*. The field vane result agrees exactly with the pressuremeter residual value but the insertion of the vane is too disruptive to see the peak strength response.

Hight et al. (1992) report that pressuremeter strength results were derived from a perfectly plastic solution. Such a methodology will inappropriately average the peak and residual strength, giving a value of approximately 50 kPa for the example.

Usually the measured data are the pressure-strain curve and the shear stress-strain curve is derived. This process can be reversed. The triaxial apparatus shear stress curves shown in Fig. 4.22 have been turned into pressuremeter curves using [4.19]. The result is given in Fig. 4.23 and compared to the pressuremeter measured data using the same reference stress. The consequences of losing the peak strength information are very obvious.

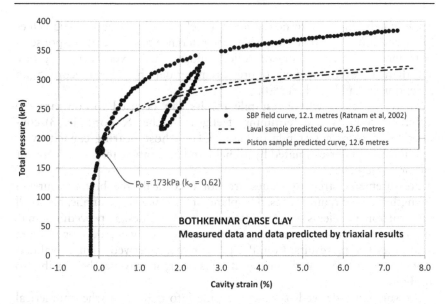

Figure 4.23 Measured and predicted field curves.

4.13 CONCLUDING REMARKS

Pressuremeters do not invariably give the highest magnitude strength parameters. Occasionally laboratory testing will give greater values, due to the non-random nature of sampling—the core has to be sensibly intact in order to be tested, whereas *in situ* all discontinuities are included. The principal difficulty is the absence of comparable data, both in terms of quality and direction of loading. The strains induced by sampling are always greater than that at which the material first yields and parts of the sample are already in a plastic condition prior to laboratory testing commencing. Re-consolidation may partially restore some of what has been lost but cannot compensate for the loss of structure.

This is a particular problem for soft clay where tolerance to sampling disturbance is lower. It is possible to use a wider range of *in situ* techniques in such material, for example cone penetrometer or vane, but the associated disturbance loses all small strain information. For a pressuremeter, disturbance is not the primary consideration because beyond the disturbed zone lies material that has yet to fail—the test mobilises the properties of this remote material regardless of the immediately surrounding state.

The pressuremeter does not directly measure strength, it is a derived parameter. Consequently it is vulnerable to a misapplied analysis. Over-simplifying the problem (such as assuming perfect plasticity in a strain softening material) will give misleading results. Historically, these data in particular have been responsible for the perception that pressuremeters significantly over-estimate strength.

Ultimately, in order to determine if a technique is giving realistic values for strength or any other parameter, the reference has to be the response of instrumented structures or controlled failures. Such experiments are rare and there are even fewer where pressuremeter data have been recorded (an exception is Ladd et al., 1980).

There are cases where test pile results can be used to compare the different methods of determining strength. Butcher & Lord (1993) consider the case of a test pile in Gault Clay whose bearing capacity was significantly under-estimated by all field and laboratory tests — except for the pressuremeter.

Pressuremeter strength values are not necessarily the best measure of strength for all circumstances. The plane of shearing is not applicable to all construction problems. Certain empirically based design procedures may implicitly require conservative parameters as they were developed from laboratory testing results. Even if a pressuremeter-derived strength value is more representative of the material behaviour, it may not be sensible to apply it.

Pressuremeter-derived parameters come into their own when numerical modelling is being used for predictive purposes. There is then no purpose to inputting conservative data.

4.14 SPECULATIVE CORRELATIONS

There are implied relationships between parameters where one part is determined by the pressuremeter, but the other must be inferred because the test does not see it. Despite the disapproval of empiricism expressed throughout this text, two such are considered here: estimating the over-consolidation ratio OCR and obtaining a drained friction angle from undrained strength. Experience with the following equations suggests they have their place but must not be relied on.

4.14.1 Overconsolidation ratio

Natural ground is Overconsolidation when its current state of vertical effective stress is less than the maximum it experienced in the past. There are a number of ratios that could be called the Overconsolidation ratio (OCR) but the particular version described here is R_p, where:

$$R_p = P'_\gamma/P'_A \qquad [4.40]$$

This is the ratio of maximum past effective stress, P'_γ, to the current effective stress, P'_A. For present purposes, P'_A is the effective overburden stress and is usually easy to estimate to an acceptable accuracy.

OCR is widely used to describe the nature of clays and as a method of relating disparate data sets so it is worthwhile to see if it can be derived from pressuremeter-determined parameters.

Wroth (1984) provides a correlation, using critical state soil mechanics nomenclature, between the undrained shear strength ratio and OCR:

$$\frac{c_u}{\sigma'_{vo}} = \frac{M}{2}\left(\frac{R_p}{r}\right)^{\Lambda} \qquad [4.41]$$

where

c_u is undrained shear strength

σ'_{vo} is the effective overburden stress

M is the frictional coefficient $= 6\sin\phi/(3 - \sin\phi)$ and is ≈ 1 for typical values of the friction angle ϕ

r is the spacing ratio between equivalent points on the isotropic consolidation line and critical state line and is ≈ 2

Λ is the plastic volumetric strain ratio and for most clays is ≈ 0.8.

The labelling here is slightly different from the published version, and Wroth is careful to specify that the undrained shear strength and friction angle are triaxial test parameters, not those from plane strain shearing. In view of the other uncertainties these are minor reservations so combining these assumptions leads to the following:

$$R_p = 2\left[2\frac{c_u}{\sigma'_{vo}}\right]^{\frac{1}{\Lambda}} \qquad [4.42]$$

Ladd et al. (1977) quote a similar expression using classical soil mechanics terminology, and this can be re-arranged to give the following:

$$R_p = \left[\left(\frac{c_u}{\sigma'_{vo}}\right)\left(\frac{1}{0.11 + 0.0037\ I_p}\right)\right]^{\frac{1}{m}} \qquad [4.43]$$

where I_P is plasticity index. Ladd et al. (1977) note that the exponent m reduces slightly with increasing OCR and has the range 0.85 to 0.75. Wroth ('84) states that $m \equiv \Lambda$.

Equ. [4.43] is related to an earlier empirical formulation offered by Skempton (1957) for natural deposits of normally consolidated clay:

$$\frac{c_u}{\sigma'_{vo}} = 0.11 + 0.0037\ I_p \qquad [4.44]$$

The values of shear strength used to develop the Skempton correlation were obtained from vane tests.

Wroth ('84) also offers the following:

$$(R_p)^\Lambda = \frac{(c_u/\sigma'_{vo})}{(c_u/\sigma'_{vo})_{nc}}$$ [4.45]

This normalises the undrained strength ratio and is independent of the frictional coefficient M and the spacing ratio r.

Based on [4.42], when values for OCR are required, it is reasonable to apply the following:

$$R_p = 2\left[2\frac{c_u}{\sigma'_{vo}}\right]^{\frac{5}{4}}$$ [4.46]

Lower-bound estimates for R_p can also be derived from p'_f/σ'_{vo}, where p'_f is the effective yield stress provided by the pressuremeter test, equating p'_y with p'_f. This makes fewer assumptions but is dependent on identifying the cavity reference pressure, p_o. A yield stress may also be identified directly from the field curve of a pre-bored test as the point of departure from a linear to non-linear trend. Ultimately, due to the stress cycling undergone by the ground as part of the pre-bored method, this may give more plausible results for R_p.

4.14.2 Inferring values for the angle of friction from undrained tests

In principle, the drained angle of internal friction, ϕ, can be derived from undrained results via the following, being the slope of the Mohr's circle relating normal stress to shear stress:

Let the mean effective normal stress be σ'_n:

$$\sigma'_n = (\sigma'_{vo} + \sigma'_{ho})/2$$ [4.47]

Then

$$\sin \phi = c_u/\sigma'_n$$ [4.48]

where c_u is undrained shear strength. This will not give credible values for ϕ using pressuremeter-derived results, especially in over-consolidated clays.

When examining vane results, Wroth ('84) gives the following relationship between the maximum shear stress and plane strain friction angle:

$$\tau_{max} = \sigma'_{ho} \sin \phi_{ps} \qquad [4.49]$$

This assumes the soil responds as a linear elastic material up to the point where the maximum shear stress is mobilised. For an undrained test in soils with a non-linear elastic/perfectly plastic shear stress/shear strain curve, Bolton & Whittle (1999) give:

$$\tau_{max} = c_u/\beta \qquad [4.50]$$

For a *drained* test in soils with a non-linear elastic shear stress/shear strain curve, it can be shown (refer to Chapter 5) that:

$$\tau_{max} = \left[\frac{p_0 \sin \phi_{ps}}{\beta + (\beta - 1)\sin \phi_{ps}} \right] \qquad [4.51]$$

Equating the right-hand side of [4.50] with the right-hand side of [4.51] and solving for $\sin \phi_{ps}$:

$$\sin \phi_{ps} = \left[\left(\frac{\sigma'_{ho}}{c_u} \right) - \left(\frac{1 - \beta}{\beta} \right) \right]^{-1} \qquad [4.52]$$

where β is the exponent of non-linearity, lies between 0.5 and 1, with 1 representing linear elasticity. If β is 1, then [4.48] and [4.52] are identical, with τ_{max} equal to c_u.

This expression generally gives sensible values for ϕ_{ps}, but the argument is speculative. Note that triaxial test determined values for friction angle, ϕ_{tx}, are not the same as ϕ_{ps}. Wroth (1984) gives $8\phi_{ps} \approx 9\phi_{tx}$.

NOTES

1 Strictly, only the plastic components of this equation are given by Windle and Wroth.
2 Ménard Limit Pressure (P_{LM}) is derived from a doubling of the initial cavity volume, when $\Delta V/V_o = 1$. In terms of *current* cavity volume this is $\Delta V/V = 0.5$. Hence, given a value for the true limit pressure, the Ménard value is calculated by deducting $c_u \log_e 2$.
3 Similar analyses were produced independently by Ladanyi (1972) and Baguelin et al. (1972).
4 We have been known to apply a straight edge to the data on the monitor screen whilst the test is being logged to read off the approximate shear stress.

5 Butler, F.G. (1995), private correspondence.
6 The nomenclature differs from the published paper to avoid symbol conflict.
7 Timoshenko & Goodier (1951) use a similar description when analysing the expansion of long cylinders.
8 Axial strain has been converted to plane shear strain in this example.

REFERENCES

Baguelin, F., Jezequel, J., Le Mee, E. & Le Mehaute, A. (1972) Expansion of a cylindrical probe in cohesive soils. *Journal of the Soil Mechanics and Foundations Division, ASCE* **98** (SM11, Paper 9377), pp. 1129–1142.

Bishop, R.F., Hill, R. & Mott, N.F. (1945) The theory of indentation and hardness tests. *Proceedings of the Physical Society* **57** (3), pp. 147–159.

Bolton, M.D. & Whittle, R.W. (1999) A non-linear elastic/perfectly plastic analysis for plane strain undrained expansion tests. *Géotechnique* **49** (1), pp. 133–141.

Butcher, A. & Lord, J. (1993) Engineering properties of the Gault Clay in and around Cambridge, UK. Proc. Int Symp. Geotechnical Engineering of Hard Soils–Soft Rocks, Athens, Greece, pp. 405–416.

Cao, L.F., Teh, C.I. & Chang, M.F. (2002) Analysis of undrained cavity expansion in elasto-plastic soils with non-linear elasticity. *International Journal for Numerical and Analytical Methods in Geomechanics* **26**, pp. 25–52.

Ferreira, R.S. & Robertson, P.K. (1992) Interpretation of undrained self-boring pressuremeter test results incorporating unloading. *Canadian Geotechnical* **29**, pp. 918–928.

Gibson, R.E. & Anderson, W.F. (1961) In situ measurement of soil properties with the pressuremeter. *Civil Engineering and Public Works Review* **56**, 658, pp. 615–618.

Hight, D., Bond, A.J. & Legge, J.D. (1992) Characterization of the Bothkennar clay: an overview. *Géotechnique* **42** (2), pp. 303–347.

Houlsby, G.T. & Carter, J.P. (1993) The effects of pressuremeter geometry on the results of tests in clay. *Géotechnique* **43** (4), pp. 567–576.

Houlsby, G. & Withers, N.J. (1988) Analysis of the cone pressuremeter test in clay. *Géotechnique* **38** (4), pp. 573–587.

Hughes, J.M.O. (1973) An instrument for in situ measurement in soft clays, PhD Thesis, University of Cambridge.

Jefferies, M.G. (1988) Determination of horizontal geostatic stress in clay with self-bored pressuremeter. *Canadian Geotechnical* **25** (3), pp. 559–573.

Ladanyi, B. (1972) In situ determination of undrained stress-strain behaviour of sensitive clays with the pressuremeter. *Canadian Geotechnical* **9**, pp. 313–319.

Ladd, C.C., Germaine, J.T., Baligh, M.M. & Lacasse, S.M. (1980) Evaluation of self-boring pressuremeter tests in Boston Blue Clay. *Interim report for Federal Highway Administration, Washington.* Ref. no. FHWA/RD-80/052

Ladd, C.C., Foott, R., Ishihara, K., Schlosser, F. & Poulos, H.G. (1977) Stress-deformation and strength characteristics: SOA report. *Proc. 9th Int. Conf. on Soil Mechanics and Foundation Eng., Tokyo* **2**, pp. 421–494.

Mair, R.J. & Wood, D.M. (1987). Pressuremeter Testing. Methods and Interpretation. Construction Industry Research and Information Association Project 335. Publ. Butterworths, London. ISBN 0-408-02434-8.

Marsland, A. & Randolph, M.F. (1977). Comparison of the results from pressuremeter tests and large in situ plate tests in London Clay. *Géotechnique* **27** (2), pp. 217–243.

Nash, D.F.T., Powell, J.J.M. & Lloyd, I.M. (1992) Initial investigations of the soft clay site at Bothkennar. *Géotechnique* **42** (2), pp. 163–181.

Palmer, A.C. (1972) Undrained plane-strain expansion of a cylindrical cavity in clay: a simple interpretation of the pressuremeter test. *Géotechnique* **22** (3), pp. 451–457.

Powell, J.J.M. (1990) A comparison of four different pressuremeters and their methods of interpretation in a stiff heavily over-consolidated clay. Proc. 3rd Int. Symp. on Pressuremeters, Oxford, April 1990, Thomas Telford, ISBN 0 7277 1556 9, pp. 287–298.

Ratnam, S., Soga, K. & Whittle, R.W. (2002). Self-boring pressuremeter permeability measurements in Bothkennar clay. *Géotechnique* **52**(1), pp. 55–60.

Shuttle, D.A. & Jefferies, M.G. (1995) A practical geometry correction for interpreting pressuremeter tests in clay. *Géotechnique* **45** (3), pp. 549–554.

Skempton, A.W. (1957) Discussion: Further data on the c/p ratio in normally consolidated clays. *Proceedings of the Institution of Civil Engineers* **7**, pp. 305–307.

Teh, C.I. (1987) An analytical study of the cone penetration test. DPhil thesis, Oxford University.

Timoshenko, S.P. & Goodier, J.N. (1951) Theory of Elasticity. 2nd Edition, McGraw-Hill Book Company, Inc.

Vardanega, P.J., & Bolton, M.D. (2013) Stiffness of clays and silts: normalizing shear modulus and shear strain. *Journal of Geotechnical and Geoenvironmental Engineering* **2013** (139), pp. 1575–1589.

Whittle, R.W. (1999) Using non-linear elasticity to obtain the engineering properties of clay. *Ground Engineering*, May, **32** (5), pp. 30–34.

Windle, D. & Wroth, C.P. (1977) The use of a Self-boring Pressuremeter to determine the undrained properties of clays. *Ground Engineering*, September, **10** (6), pp. 37–46.

Wood, D.M. & Wroth, C.P. (1977) Some laboratory experiments relating to the results of pressuremeter tests. *Géotechnique* **27** (2), pp. 181–201.

Wroth, C.P. (1978) Cambridge in-situ probe. Proc. Symp. on site exploration in soft ground using in situ techniques, Virginia, May 1978. FHWA-TS-80-202, pp. 97–135.

Wroth, C.P. (1984) The interpretation of in situ soil tests. *Twenty Fourth Rankine Lecture, Géotechnique* **34** (4), pp. 449–489.

Chapter 5

Analyses for Drained Strength

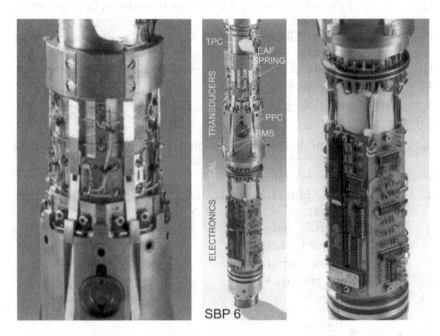

Figure 5.0 Inside the six-arm self-boring pressuremeter.

5.1 NOTATION

r	is current cavity radius. r_o is the initial radius of the cavity.
R	Radius of the elastic/plastic boundary
ε_c	is current cavity strain, measured by the pressuremeter and is $(r - r_o)/r$.
ε_t	is tangential strain and is $(r_o - r)/r_o$ (used in Manassero analysis).
ε_{ip}	is circumferential yield strain when expanding

DOI: 10.1201/9781003200680-5

ε_{ipu}	is circumferential yield strain when contracting
ε_{mx}	is the maximum cavity strain at the end of loading
p'_c	is effective pressure at the cavity wall and is measured by the pressuremeter
p'_o	is effective cavity reference pressure (total reference pressure when prime is omitted)
p'_f	Radial stress at the cavity wall at first yield
p'_{fu}	Radial stress at the cavity wall when yielding in contraction
p'_{mx}	is the maximum effective cavity pressure at the end of loading
p'_{lim}	Effective limit pressure
e	Void ratio
G	is shear modulus, G_s is secant shear modulus, G_t is tangential shear modulus
G_{load}	Shear modulus at first yield, G_{unload} is shear modulus at yield in contraction
G_{base}	Constant of the stiffness stress dependency relationship
K_a	is $(1 - \sin\phi')/(1 + \sin\phi')$
K_a^{cv}	is $(1 - \sin\phi'_{cv})/(1 + \sin\phi'_{cv})$
K_p^{cv}	is $(1 + \sin\phi'_{cv})/(1 - \sin\phi'_{cv}) = 1/K_a^{cv}$
S, (S_{ul})	Slope of the loading (unloading) response when the field data are plotted on log – log scales
v	is Poisson's ratio
ϕ'	is angle of shearing resistance. ϕ'_{cv} is the friction angle when the material is shearing at constant volume.
ψ	is dilation angle
c'	is drained cohesion
σ	is stress, suffix r for radial stress, θ for circumferential stress
σ'_{av}	Mean effective stress
M	is $\dfrac{(1 + \sin\psi)}{(1 - \sin\psi)}$
N	is $\dfrac{(1 + \sin\phi)}{(1 - \sin\phi)}$
τ	is shear stress
γ	Shear strain
α	Shear stress constant (Bolton & Whittle, 1999). α_{ref} denotes α at a particular σ'_{av}. α_r and α_{ru} are radial stress constants for loading and unloading, respectively.
β	Exponent of non-linearity (Bolton & Whittle, 1999)
ξ	State parameter
λ, Γ	The slope and intercept of the critical state line
J	Exponent of stress dependency (Janbu, 1963)
A	Area, specifically constant area ratio $\Delta A/A = \Delta V/V$ where V is volume

A, B	Substitution terms applied within the Houlsby et al. (1986) description
c_c	Small elastic constant
I_d	Relative density
A_m, B_m, C_m, D_m	Substitution terms applied within the Manassero (1989) description

The following notation is specific to the Carter et al., 1986 *solution:*

χ	is given by $[(1 - \nu) - \nu(M + N) + (1 - \nu)MN]/MN$
Z	is $2\chi MN/(N + M)$
D	is $M(2 + Z)/(M + 1)$
F	is $- ZN/(N - 1)$
H	is $1 - D - F$

5.2 INTRODUCTION

When a cavity expansion is a drained event, strength varies with the normal stress (Fig. 5.1).

If the shear stress/strain response is assumed to be elastic/perfectly plastic then following yield the shear stress changes at a constant stress ratio and at the peak angle of shearing resistance, ϕ'.

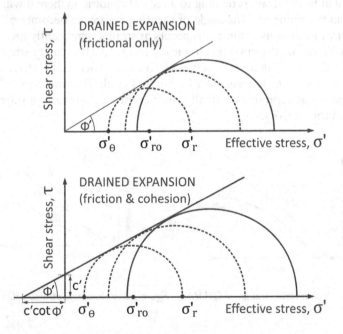

Figure 5.1 Mohr's circles for drained failure.

Compared to the undrained situation, two additional parameters have to be decided or given that are not explicitly determined by the cavity expansion test:

- Ambient water pressure, u_0.
- Residual or constant volume friction angle, ϕ'_{cv}.

u_0 is required to allow the total stress measurements to be converted to effective stress. It can often be recognised from the closing stages of the pressuremeter field curve when the membrane collapses. However, drilling procedures frequently introduce water into the borehole and this may give a misleading impression of the ambient level.

ϕ'_{cv} can be roughly estimated from direct observation of the material and sometimes from the observed response at large expansion when the dilation tends towards zero.

The major difficulty with the drained interpretation is accounting for the volumetric strains that occur as the material dilates or compresses. This is resolved by Hughes et al. (1977), by incorporating Rowes' dilatancy rule (1962) into the solution. This method for calculating the volumetric strain contribution has been included in all subsequent analyses for drained cavity expansion and contraction.

The assumption that the material deforms at a constant stress ratio is only valid for a relatively small strain deformation. The actual response of the material at larger strain is tending to a critical condition where it will deform at constant volume and the angle of shearing resistance becomes ϕ'_{cv}. Most closed-form solutions assume this occurs at strains significantly greater than those achieved in a normal test (typically less than 10% cavity strain).

Figs. 5.2 and 5.3 sketch the actual behaviour of frictional material and the idealised representation assumed by Hughes et al. The intercept "c" in the idealised image represents a small elastic constant that was considered negligible in the analysis.

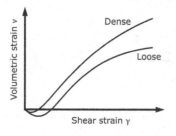

Figure 5.2 Volumetric strains—ACTUAL.

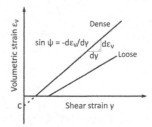

Figure 5.3 Volumetric strains—IDEALISED.

One solution that does not require this assumption is the Manassero, (1989) numerical analysis. It is sensitive to the smoothness of the data, but under ideal circumstances follows the actual material response and places no restrictions on the current internal angle of friction.

As with the Palmer analysis for the undrained case, the primary issue with the numerical approach is the requirement for certainty in the starting condition. Closed-form solutions applied as a model can be arranged to discover the initial state, so despite their limitations remain the preferred choice for analysing the shearing of frictional material.

For finding the effective limit pressure, p'_{lim}, of the material, both closed-form and numerical solutions can be used.

5.2.1 Drained cohesion

Hughes et al. assume that the material response is purely frictional and ignore the possibility of drained cohesion, c'. This leads to a convenient graphical solution for the test (Fig. 5.4) where the gradient of the experimental data in log-log space is a function of the frictional properties only and is readily expressed as angles of internal friction and dilation.

It is straightforward to modify this solution for the additional contribution of drained cohesion but it is not easy to isolate c' in the presence of friction.

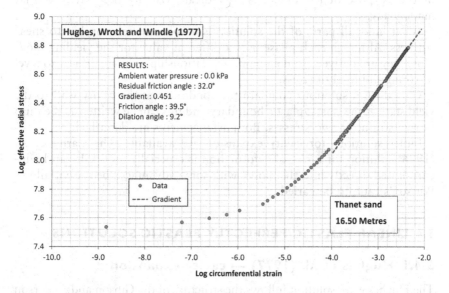

Figure 5.4 An example of the Hughes et al., (1977) construction.

If the analysis is used as the basis of a curve comparison model, then it is generally possible to find the optimum fit of field data to the model that allows cohesion to be quantified. Compared to the undrained case, the model is significantly less constrained and requires some judgement to interpret.

5.3 DRAINED CURVE MODELLING

A model for the drained pressuremeter test can be constructed from two closed-form solutions for drained cylindrical cavity expansion (Hughes et al., 1977) and contraction (Houlsby et al., 1986) in a purely frictional material. In both cases, the solutions assume the material response is elastic/perfectly plastic (Gibson & Anderson, 1961).

The model is easily adapted to incorporate cohesion using the approach suggested by Carter et al. (1986). Additionally, the yield condition can be re-defined by a non-linear relationship based on a power law, adapting the method outlined by Bolton and Whittle (1999). The model will then describe the drained pressuremeter test in $c' - \phi'$ material in a manner analogous to the undrained model described in the previous chapter.

The two phases of the model are largely uncoupled for strength, so separate values for the angle of shearing resistance are required for the loading compared to the unloading. The adoption of a single value for each case is an inadequate description of the material response, but part of the purpose of the model is to indicate visually the extent of that inadequacy.

What links all parts of the solution is a coherent relationship to shear stiffness. Although each phase uses a different value for the yielding modulus, they are derived from a common relationship linking stiffness to level of strain and level of stress.

Closed-form solutions compare the conditions at the cavity wall to the yield state at the elastic/plastic boundary and this can be written as the ratio of two strains or two stresses. Both arrangements are described.

Other solutions for cavity expansion are available. The Carter et al. (1986) solution for a $c' - \phi'$ deformation is especially useful. It can be incorporated directly into the model with some advantages because it allows Poisson's ratio to be varied.

5.4 LINEAR ELASTIC/PERFECTLY PLASTIC SOLUTIONS

5.4.1 Hughes et al. (1977) — cavity expansion

The Hughes et al. solution follows the structure of the Gibson and Anderson (1961) solution, but takes account of the development of volumetric strains using Rowe's dilatancy rule (1962). This relates the peak angle of internal friction, ϕ', to the critical state angle, ϕ'_{cv} (when the material is deforming at constant volume) via a dilation angle, ψ. This can be written as:

Rowe, (1962):

$$\frac{(1 + \sin \phi')}{(1 - \sin \phi')} = \frac{(1 + \sin \phi'_{cv})}{(1 - \sin \phi'_{cv})} \frac{(1 + \sin \psi)}{(1 - \sin \psi)} \qquad [5.1]$$

Equations [5.2] to [5.6] set out the solution but in a different form to that given in the published paper.

The particular advantage of the Hughes solution can be seen by inspecting [5.5] or [5.6]. The slope of a plot of p'_c against ε_c on log scales predicts a near linear trend, the gradient of which is the exponent terms. This in turn can be re-arranged to give the internal angle of friction and dilation using [5.8] and [5.9]. Fig. 5.4 is an example.

Linear elastic loading:

$$p'_c - p'_0 = 2G_{load}\varepsilon_c \qquad [5.2]$$

Yield stress:

$$p'_c = p'_f = p'_0(1 + \sin\phi') \qquad [5.3]$$

The yield strain depends on p'_f and the appropriate shear modulus:

$$\varepsilon_c = \varepsilon_{ip} = \frac{p'_f - p'_0}{2G_{load}} \qquad [5.4]$$

Plastic loading, solved for radial stress:

$$p'_c = p'_f\left[\frac{\varepsilon_c}{\varepsilon_{ip}(1 + \sin\psi)} + \left(\frac{\sin\psi}{(1 + \sin\psi)}\right)\right]^S \qquad [5.5]$$

Alternatively, solved for cavity strain:

$$\varepsilon_c = \varepsilon_{ip}(1 + \sin\psi)\left[\left(\frac{p'_c}{p'_f}\right)^{\frac{1}{S}} - \left(\frac{\sin\psi}{(1 + \sin\psi)}\right)\right] \qquad [5.6]$$

The exponent S is defined as:

$$S = \frac{(1 + \sin\psi)\sin\phi'}{1 + \sin\phi'} \qquad [5.7]$$

Sin ϕ' is derived from S:

$$\sin \phi' = \frac{s}{1 + (s - 1)\sin \phi'_{cv}} \qquad [5.8]$$

Sin ψ can also be derived from S:

$$\sin \psi = s + (s - 1)\sin \phi'_{cv} \qquad [5.9]$$

Cohesion is introduced by raising all stresses by the quantity $c'\cot\phi'$ and modifying the yield stress criterion. Note that redefining p'_f also modifies the yield strain, ε_{ip}, in [5.4]:

Yield stress:

$$p'_c = p'_f = p'_0(1 + \sin\phi') + c'\cos\phi' \qquad [5.10]$$

[5.6] becomes

$$\varepsilon_c = \varepsilon_{ip}(1 + \sin\psi)\left[\left(\frac{p'_c + c'\cot\phi'}{p'_f + c'\cot\phi'}\right)^{\frac{1}{s}} - \left(\frac{\sin\psi}{(1 + \sin\psi)}\right)\right] \qquad [5.11]$$

[5.5] becomes

$$p'_c = (p'_f + c'\cot\phi')\left[\frac{\varepsilon_c}{\varepsilon_{ip}(1 + \sin\psi)} + \left(\frac{\sin\psi}{(1 + \sin\psi)}\right)\right]^S - c'\cot\phi' \qquad [5.12]$$

Non-linearity will be introduced later but essentially is just ε_{ip} and p'_f redefined; the structures of [5.11] and [5.12] remain the same. Equations [5.10] to [5.12] are the solution for a $c' - \phi'$ material but will give the purely frictional result when cohesion is zero.

5.4.2 Houlsby et al. (1986) — cavity contraction

The contraction phase of the pressuremeter tests commences with the cavity at a maximum strain, ε_{mx}, and effective radial stress, p'_{mx}:

Elastic unloading:

$$p'_{mx} - p'_c = 2G_{unload}(\varepsilon_{mx} - \varepsilon_c) \qquad [5.13]$$

Yield stress in contraction:

$$p'_c = p'_{fu} = p'_{mx}\left(1 - \frac{1}{N}\right)$$

[5.14]

Yield strain in contraction:

$$\varepsilon_c = \varepsilon_{ipu} = \frac{(p'_{mx} - p'_{fu})}{2G_{unload}}$$

[5.15]

Plastic unloading, solved for stress:

$$p'_c = p'_{fu}\left[A\left(\frac{\varepsilon_{mx} - \varepsilon_c}{\varepsilon_{ipu}}\right) - B\right]^{S_{ul}}$$

[5.16]

where

$$
\begin{aligned}
A &= \frac{NM + 1}{N + 1} = \frac{1 + \sin\phi'\sin\psi}{1 - \sin\psi} \\
B &= \frac{NM - N}{N + 1} = \frac{\sin\psi(1 + \sin\phi')}{1 - \sin\psi} \\
S_{ul} &= \frac{(1 + N)(1 - N)}{1 + NM} = \frac{-2\sin\phi'(1 - \sin\psi)}{(1 - \sin\phi')(1 + \sin\phi'\sin\psi)}
\end{aligned}
$$

[5.17]

Plastic unloading, solved for strain:

$$\varepsilon_c = \varepsilon_{mx} - \frac{\varepsilon_{ipu}}{A}\left[\left(\frac{p'_c}{p'_{fu}}\right)^{1/S_{ul}} + B\right]$$

[5.18]

As with the loading equations, drained cohesion is added to the contraction solution by increasing all stresses by $c'\cot\phi'$ and redefining the yield definitions. [5.18] becomes:

$$\varepsilon_c = \varepsilon_{mx} - \left(\frac{\varepsilon_{ipu}}{A}\right)\left[\left(\frac{p'_c + c'\cot\phi'}{p'_{fu} + c'\cot\phi'}\right)^{1/S_{ul}} + B\right]$$

[5.19]

[5.16] becomes:

$$p'_c = (p'_{fu} + c'\cot\phi')\left[A\left(\frac{\varepsilon_{mx} - \varepsilon_c}{\varepsilon_{ipu}}\right) - B\right]^{S_{ul}} - c'\cot\phi'$$

[5.20]

Yield stress in contraction:

$$p'_{fu} = p'_{mx} - \frac{2\,(p'_{mx}\,\sin\phi' + c'\cos\phi')}{1 + \sin\phi'} \qquad [5.21]$$

Re-defining the yield stress modifies the yield strain, ε_{ipu}, calculation in [5.15]. Houlsby et al. apply their solution as a curve modelling procedure. Withers et al. (1989) show that [5.16] can be written in the following form:

$$\log_e\left(\frac{p'_c}{p'_{mx}}\right) = s_{ul}\,\log_e\left[\left(\frac{\varepsilon_{mx} - \varepsilon_c}{1 + \varepsilon_{mx}}\right) + c_c\right] + \text{constant} \qquad [5.22]$$

c_c is a negligibly small elastic constant. It follows that a plot of $\frac{p'_c}{p'_{mx}}$ against $\left(\frac{\varepsilon_{mx} - \varepsilon_c}{1 + \varepsilon_{mx}}\right)$ on log scales gives a trend whose final slope is the unloading exponent, s_{ul} (Fig. 5.5). This permits the peak friction and dilation angles to be derived:

$$K_a = \sqrt{\left[\frac{s_{ul}K_a^{cv} + 1}{1 - s_{ul}}\right]}$$

$$\sin\phi' = \frac{1 - K_a}{1 + K_a} \qquad [5.23]$$

$$\sin\psi = \frac{1 - K_a/K_a^{cv}}{1 + K_a/K_a^{cv}}$$

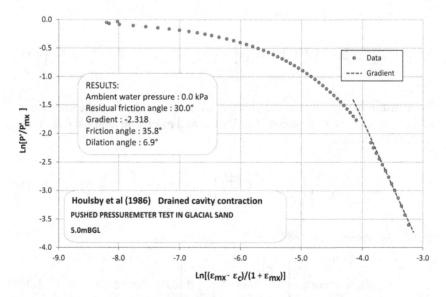

Figure 5.5 An example of the Houlsby et al. solution for cavity contraction.

5.5 INCORPORATING NON-LINEARITY

5.5.1 Non-linear yield

At yield, the material reaches the failure surface for the first time and the stress ratio, σ'_r/σ'_θ, becomes constant. Assuming that the material response up to yield can be described by a power law in a manner analogous to the undrained configuration (Bolton & Whittle, 1999), the following is obtained:

$$\sigma'_r = p'_f = p'_o + \frac{\tau}{\beta}$$
$$\sigma'_\theta = \sigma'_r - 2\tau \tag{5.24}$$

Substituting these definitions into the stress ratio N leads to the purely frictional results for non-linear yield (refer to Appendix A where these relationships are developed in more detail):

For first yield:

$$\sigma'_r = p'_f = p'_o + \left(\frac{p'_o \, \sin\phi'}{\beta(1 + \sin\phi') - \sin\phi'} \right) \tag{5.25}$$

For yield in contraction:

$$\sigma'_r = p'_{fu} = p'_{mx} - \left(\frac{2p'_{mx} \, \sin\phi'}{\beta(1 - \sin\phi') + 2\sin\phi'} \right) \tag{5.26}$$

Cohesion can be introduced into these definitions in the manner already described:

For first yield:

$$\sigma'_r = p'_f = p'_o + \left(\frac{(p'_o \, \sin\phi' + c'\cos\phi')}{\beta(1 + \sin\phi') - \sin\phi'} \right) \tag{5.27}$$

For yield in contraction:

$$\sigma'_r = p'_{fu} = p'_{mx} - \left(\frac{2(p'_{mx} \sin\phi' - c'\cos\phi')}{\beta(1 - \sin\phi') + 2\sin\phi'} \right) \tag{5.28}$$

5.5.2 Non-linear stress and strain below yield

The mean effective stress, σ'_{av}, at the cavity wall will be highest at the end of the loading and this condition will apply throughout most of the cavity contraction. Even if the material response is treated as linear elastic/perfectly plastic, it is still necessary to define two values for shear modulus: G_{load} and G_{unload}. These values could be derived from unload/reload cycles taken at strategic points during the expansion phase.

A more elegant data match can be achieved by extracting the radial stress component of stiffness from non-linear modulus that has been adjusted for stress level. Fig. 5.6 shows a field curve from a self-bored test in a dense sand that includes five unload/reload cycles. For each cycle, the Bolton and Whittle procedure is carried out to find the two power law parameters that describe the *strain*-dependent secant shear modulus. These are used to find the shear modulus at a single representative strain, here selected to be 0.3% shear strain (a typical yield strain for a sand). The results are given in Table 5.1 and plotted in Fig. 5.7 (Table 5.2).

A shear modulus (for one value of shear strain γ) is plotted against the current mean effective stress, σ'_{av}, calculated at the commencement of the cycle. The results follow a power curve, as predicted by Janbu (1963). The purpose is to identify G_{load} from the relevant mean effective stress, which in this case will be derived from the yield stress of the material. p' is p'_f for G_{load} and p'_{mx} for G_{unload}. As described in Chapter 2, the power law constituents of these modulus values can be used to find the stiffness response in radial stress/cavity strain space and hence provide curve fit data for the parts of the field curve prior to yield (in either sense). If α_r is α for radial stress/cavity strain, then the following applies:

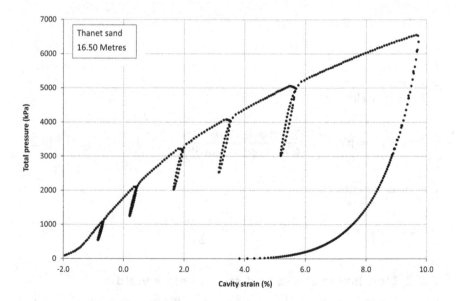

Figure 5.6 SBP test in dense sand with five cycles.

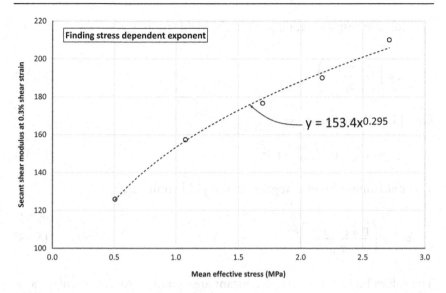

Figure 5.7 Stress dependency of stiffness.

Table 5.1 Modulus results for example test

Cycle	β	α (MPa)	p′ (kPa)	σ_av (MPa)	G_0.3% (MPa)
1	0.866	57.834	1075	0.506	126
2	0.864	71.449	2102	1.076	157
3	0.860	78.323	3216	1.693	177
4	0.837	73.736	4077	2.171	190
5	0.831	78.694	5025	2.712	210

Table 5.2 Stress-dependent results

Description	Value	Notation
Stress constant (MPa)	153.4	G_{base}
Stress exponent	0.295	J
R^2	0.995	

$$G_{load} = \alpha_{ref}\gamma^{\beta-1} = G_{base}(\sigma'_{av})^{J}$$

$$\therefore \alpha_{ref} = \frac{G_{base}(\sigma'_{av})^{J}}{\gamma^{\beta-1}} \tag{5.29}$$

$$\alpha_r = 10^{\left(\log\frac{\alpha}{\beta}+\beta\log 2\right)} \tag{5.30}$$

Equ. [5.2] is replaced by:

$$p'_c - p'_0 = \alpha_r(\varepsilon_c)^{\beta} \tag{5.31}$$

This loading response applies up to the yield strain, ε_{ip}:

$$\varepsilon_{ip} = \left(\frac{p'_f - p'_o}{\alpha_r} \right)^{\frac{1}{\beta}} \qquad\qquad [5.32]$$

Equ. [5.13] is replaced by:

$$p'_{mx} - p'_c = \alpha_{ru}(\varepsilon_{mx} - \varepsilon_c)^\beta \qquad\qquad [5.33]$$

This unloading response applies up the yield strain, ε_{ipu}:

$$\varepsilon_{ipu} = \left(\frac{p'_{mx} - p'_{fu}}{\alpha_{ru}} \right)^{\frac{1}{\beta}} \qquad\qquad [5.34]$$

The values for the shear stress constant are optimised for a particular value of shear strain γ (again, this will be different for the load and unload case), but they are insensitive to small changes in the choice. Within any implementation of the model, an iterative loop can be used to re-calculate the various stiffness parameters, closing in on the optimum value for α as the yield shear strain becomes modified. Because yield strain in contraction is greater than when first loading, it is not necessarily the case that G_{unload} is significantly larger than G_{load}, but in general $\alpha_r \neq \alpha_{ru}$.

An example of applying the model with [5.31] and [5.33] is shown in Fig. 5.8. An example in similar material but with some cohesion is shown

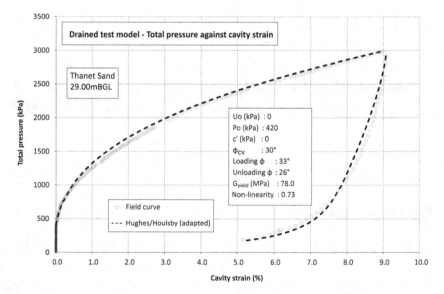

Figure 5.8 Curve fitting a drained test with no cohesion.

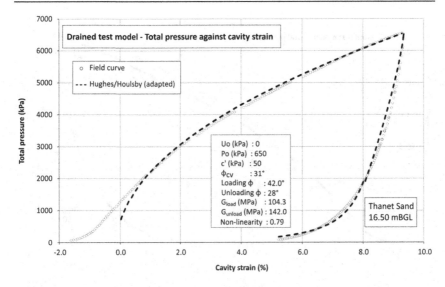

Figure 5.9 Curve fitting a drained test with cohesion.

in Fig. 5.9. This is the same test shown in Fig. 5.6, the multiple cycles allowing the shear stiffness to be closely specified.

5.5.3 Sensitivity of the model to parameter variation

Because of the way the drained model is arranged, the loading is far more affected by variations to the parameters than the contraction. The examples that follow demonstrate the effects of altering the strength and reference stress parameters. The test is that shown in Fig. 5.9. Stiffness is defined by the multiple measurements of unload/reload response and is considered fixed.

In the case of the cavity reference pressure, p_o (Fig. 5.10), it is only the loading that sees the consequences, the contraction phase is defined by the maximum pressure reached at the end of loading. Small changes to the reference stress have an effect that is immediately obvious. This is a desirable characteristic.

Changing the drained cohesion is less obvious, but affects the whole curve. Fig. 5.11 is an example. This is slightly misleading because the estimated cohesion is not very large prior to adjustment. The effects are similar to the cavity reference pressure but reduce in significance at higher expansion.

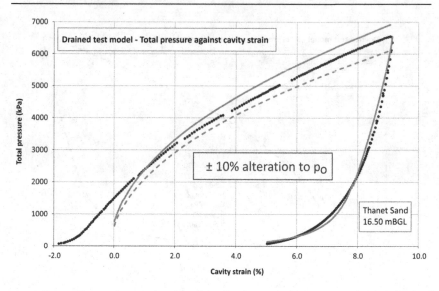

Figure 5.10 Variation of cavity reference pressure.

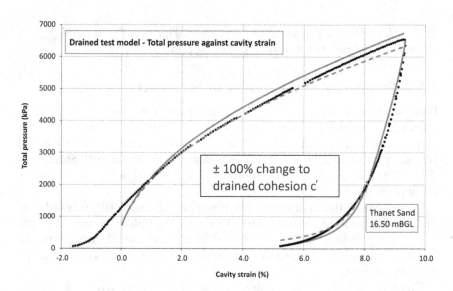

Figure 5.11 Varying the drained cohesion.

Changing the friction angle has a very marked effect. The example (Fig. 5.12) modifies the loading and contraction values by ±10% and the consequences are noticeable on both parts of the curve. For the loading, it has a similar effect to altering the cavity reference pressure.

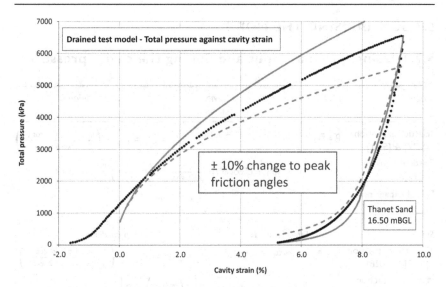

Figure 5.12 Varying the friction angle.

5.6 USE WITH ROCK

Fig. 5.13 is an example of the model applied to a test in very weak mudstone. This is a pre-bored test and is lacking data in the early stages but the fit is reasonably persuasive. In general it will only be possible to use the model for rocks if there is negligible tensile strength.

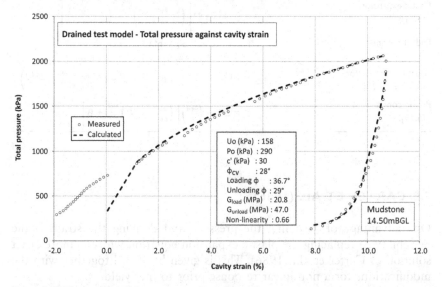

Figure 5.13 Drained model in very weak mudstone.

5.7 SUMMARISING THE MODEL

5.7.1 Using strain as input and finding the cavity pressure

Stage	Effective stress	Cavity strain	Stop condition
"Elastic" expansion [5.27] & [5.31]:	$p'_c = p'_0 + \alpha_r(\varepsilon_c)^\beta$	$(r - r_0)/r$	$p'_c = p'_f = p'_0 + \left(\frac{(p'_0 \sin\phi' + c' \cos\phi')}{\beta(1 + \sin\phi') - \sin\phi'} \right)$
Plastic expansion [5.12]:	$p'_c = (p'_f + c' \cot\phi') \left[\frac{\varepsilon_c}{\varepsilon_{ip}(1 + \sin\psi)} + \left(\frac{\sin\psi}{(1 + \sin\psi)} \right) \right]^S - c' \cot\phi'$		
End of expansion		$p'_c = p'_{mx}$ $\varepsilon_c = \varepsilon_{mx}$	

Stage	Effective stress		Stop condition
"Elastic" contraction [5.33] & [5.28]	$p'_c = p'_{mx} - \alpha_{ru}(\varepsilon_{mx} - \varepsilon_c)^\beta$	$p'_c = p'_{fu} = p'_{mx} - \left(\frac{2(p'_{mx} \sin\phi' - c' \cos\phi')}{\beta(1 - \sin\phi') + 2\sin\phi'} \right)$	
Plastic contraction [5.20]:	$p'_c = (p'_{fu} + c' \cot\phi') \left[A\left(\frac{\varepsilon_{mx} - \varepsilon_c}{\varepsilon_{ipu}} \right) - B \right]^{s_{ul}} - c' \cot\phi'$		

5.7.2 Using pressure as input and finding the cavity strain

Stage	Cavity strain	Stop condition
"Elastic" expansion: [5.32] & [5.27]:	$\varepsilon_c = \left(\frac{p'_c - p'_o}{\alpha_r} \right)^{\frac{1}{\beta}}$	$p'_c = p'_f = p'_o + \left(\frac{(p'_o \sin\phi' + c' \cos\phi')}{\beta(1 + \sin\phi') - \sin\phi'} \right)$
Plastic expansion [5.11]:	$\varepsilon_c = \varepsilon_{ip}(1 + \sin\psi) \left[\left(\frac{p'_c + c' \cot\phi'}{p'_f + c' \cot\phi'} \right)^{\frac{1}{s}} - \left(\frac{\sin\psi}{(1 + \sin\psi)} \right) \right]$	
End of expansion		$p'_c = p'_{mx}$ $\varepsilon_c = \varepsilon_{mx}$
"Elastic" contraction [5.34] & [5.28]	$\varepsilon_c = \varepsilon_{mx} - \left(\frac{p'_{mx} - p'_c}{\alpha_{ru}} \right)^{\frac{1}{\beta}}$	$p'_c = p'_{fu} = p'_{mx} - \left(\frac{2(p'_{mx} \sin\phi' - c' \cos\phi')}{\beta(1 - \sin\phi') + 2\sin\phi'} \right)$
Plastic contraction [5.19]:	$\varepsilon_c = \varepsilon_{mx} - \left(\frac{\varepsilon_{ipu}}{A} \right) \left[\left(\frac{p'_c + c' \cot\phi'}{p'_{fu} + c' \cot\phi'} \right)^{1/s_{ul}} + B \right]$	

5.8 CARTER ET AL. (1986)

One advantage of assuming the pressure and deriving the strain is the possibility of replacing the plastic expansion equation with the more exact solution of Carter et al., (1986). This is given by [5.35] together with the modifications for a non-linear response prior to first yield.

Carter et al. (1986):

$$\varepsilon_c = \varepsilon_{ip} \left[D\left(\frac{p'_c + c'\cot\phi'}{p'_f + c'\cot\phi'}\right)^{\frac{1}{s}} + F\left(\frac{p'_c + c'\cot\phi'}{p'_f + c'\cot\phi'}\right) + H \right] \qquad [5.35]$$

where

$$p'_f = p'_o + p'_o \sin\phi' + c'\cos\phi'$$

$$\varepsilon_{ip} = \frac{p'_f - p'_o}{2G}$$

$$D = M(2 + Z)/(M + 1)$$

$$F = -ZN/(N - 1)$$

$$Z = 2\chi MN/(N + M)$$

$$H = 1 - D - F$$

$$\chi = [(1 - \nu) - \nu(M + N) + (1 - \nu)MN]/MN$$

This may be modified for non-linearity by replacing p'_f and ε_{ip} with expressions already given:

$$p'_f = p'_o + \left[\frac{p'_o \sin\phi' + c'\cos\phi'}{\beta(1 + \sin\phi') - \sin\phi'}\right]. \qquad [5.27]$$

$$\varepsilon_{ip} = \left(\frac{p'_f - p'_o}{\alpha_r}\right)^{\frac{1}{\beta}} \qquad [5.32]$$

5.8.1 Elastic strains in the plastic region

Because the loaded area does not deform at constant volume, a small component of the shear strains after yield are due to elastic compression. The Hughes et al. (1977) solution ignores these elastic strains created in the plastic region. In effect, the solution is assuming Poisson's ratio, ν, is 0.5.

This can be problematic for certain combinations of parameters, in particular when dilation or contraction is large and stiffness is low. Fig. 5.14

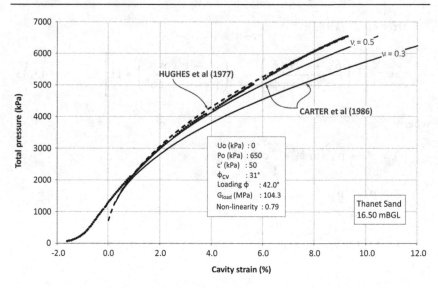

Figure 5.14 Hughes and Carter compared.

illustrates the difference between the two solutions for the test example given in Fig. 5.9 using the same parameter set. The difference increases as Poisson's ratio reduces. With all other parameters unchanged, the Carter solution predicts higher values for the angle of internal friction — in this example at least 2° more (for a v of 0.5) and potentially 5° at lower values of v.

It is not necessarily the case that applying the more complex solution is the better option. Poisson's ratio is a property the pressuremeter test does not measure; it has to be estimated. Moreover, it is not fixed but increases with shear strain. Towards the end of loading it is approaching the constant volume state. The effect is that for the early stages of the test the Carter solution is the more realistic representation. At large strain it is more likely to mislead than the Hughes solution.

Both solutions assume that the internal angle of friction is constant (over the strain range of a typical test). The Hughes solution is easily reduced to its constituent parts (refer to Appendix A), allowing this requirement to be tested and quantified. The Carter solution is a curve-matching procedure. If used, it may be helpful to first plot the field data on log scales to find an approximate value for the gradient S.

These are engineering decisions. Both methods assume perfect plasticity and can only be "right" for a limited strain range. Hughes is essentially conservative—it will always under-estimate the internal angle of friction.

5.8.2 Deriving the limit pressure from Carter et al. (1986)

If the shear strains are sufficiently large (probably greater than 20%), then eventually there will be a coordinate of stress and strain after which the material will deform at constant volume. From this point on, there is no operational difference between the deformation of a drained sand and an undrained clay. It follows that an ultimate or limit stress can be calculated where the cavity will expand indefinitely if that pressure is maintained.

Despite being a small strain solution, it is possible to use the Carter et al. solution in its adapted form to discover the limit pressure of an infinitely large expansion. At the limit state, the ratio R/r of the elastic-plastic boundary to the current cavity size reaches a constant condition, which can be written:

$$\frac{1}{\varepsilon_{ip}} = \left[(2 + Z) \left(\frac{p'_{lim}}{p'_f} \right)^{1/s} - Z \left(\frac{p'_{lim}}{p'_f} \right) \right] \qquad [5.36]$$

where P_{lim} is limit pressure. To utilise, start with knowing the elastic yield strain, ε_{ip}. Now guess the ratio p'_{lim}/p'_f and apply [5.36] within an iterative procedure, modifying the ratio until the known value of ε_{ip} is obtained. Once the ratio has been identified, the product of the ratio and the effective yield stress, p'_f, give the effective limit pressure, p'_{lim}.

Yu & Houlsby (1991) state that this solution can only be considered approximate.

5.9 MANASSERO (1989)

5.9.1 Introduction

The Manassero analysis is a numerical procedure that makes the same assumptions about the deformation mode of a frictional material sheared under drained conditions as Hughes et al. (1977). However, Rowe's dilatancy relationship (Rowe, 1962) is applied as a flow rule so there is no requirement to assume deformation at a single value of friction angle. Hence, the solution permits the variable degrees of volume change during shear.

The advantage is a more representative description of the shear stress/shear strain response. The disadvantage is that the numerical method is very sensitive to minor fluctuations in the measured data and is limited by the spacing between data points. Manassero suggests that the measured data be fitted with a polynomial function prior to implementing the numerical calculations but it is often difficult to preserve the nuances of the true ground response with such arrangements and it may be better to live with the data instability. Where possible, the test itself can be run in a manner that minimises the data uncertainty, using controlled leaks to alter the internal pressure rather than electronic valves. For smoother parts of the test such as unload/reload cycles it is the data spacing that is the issue. Curve fitting the reloading data with a power function to get the response in radial stress/cavity strain space (this is set out in Chapter 2) means that the Manassero equations can be used to obtain the equivalent trend in shear stress–shear strain space.

5.9.2 Derivation

At the borehole wall, ε_t and σ'_r are obtained directly from measurements the pressuremeter provides of radial displacement and total pressure p_c.[1] This becomes effective stress, p_c', when the ambient pore water pressure is deducted. The solution finds the radial strain, ε_r, by carrying out a series of finite difference calculations using the current gradient of the measured field curve. ε_r at a point (i) is obtained as follows:

$$\varepsilon_r^{(i)} = A_m - B_m + C_m + D_m. \tag{5.37}$$

where

$$A_m = \frac{p_c'^{(i)} \left[\varepsilon_t^{(i-1)} + k_a^{cv} \varepsilon_r^{(i-1)} \right]}{2 \left[p_c'^{(i)} \left(1 + k_a^{cv} \right) - p_c'^{(i-1)} \right]}$$

$$B_m = \frac{p_c'^{(i-1)} \varepsilon_t^{(i)}}{2 \left[p_c'^{(i)} \left(1 + k_a^{cv} \right) - p_c'^{(i-1)} \right]}$$

$$C_m = \frac{p_c'^{(i)} \left[\varepsilon_t^{(i-1)} - \varepsilon_r^{(i-1)} \right]}{2 k_a^{cv} p_c'^{(i-1)}}$$

$$D_m = \frac{p_c'^{(i-1)} \left[\varepsilon_r^{(i-1)} \left(1 + k_a^{cv} \right) - \varepsilon_t^{(i)} \right]}{2 k_a^{cv} p_c'^{(i-1)}}$$

Because the expansion of the cavity starts from zero radial strain, the first interval is defined and allows the series of linked calculations to be implemented.

Once ε_r, is known, volumetric strain ε_v and shear strain γ are calculated from the sum and difference of ε_r and ε_c. Additionally, the principal stress relationship can be obtained through:

$$\frac{\sigma_r}{\sigma_c} = k_p^{cv} \frac{\Delta \varepsilon_t}{\Delta \varepsilon_R} \tag{5.38}$$

This allows the circumferential stress σ_c to be obtained and this in turn permits shear stress τ to be derived:

$$\tau = \frac{\sigma_r - \sigma_c}{2} \qquad (5.39]$$

The current friction angle is given by:

$$\sin \phi' = \frac{\sigma_r - \sigma_c}{\sigma_r + \sigma_c} \qquad [5.40]$$

An example of the Manassero solution applied to an ideal self-bored pressuremeter test is given in Fig. 5.15. A more representative example from a test in the field is given in Fig. 5.16. The "ideal" example uses given data in the Manassero paper and was conducted in a calibration chamber where the material was placed with the probe already in position. No self-boring was actually carried out, unlike the field test.

The slope values in these figures are the sines of the dilation angles (see Fig. 5.3). At the end of the Manassero chamber test, the material is showing a curved response as it tends towards constant volume deformation. The field test in alluvial sand does not show this. However, it is normal in dense sand to see some indication of dilation declining at shear strains in excess of 15%.

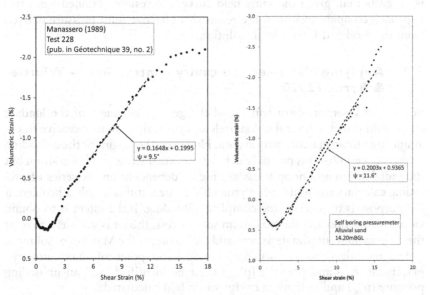

Figure 5.15 Volumetric strain vs shear strain.   Figure 5.16 Volumetric strain vs shear strain—field data.

Both figures give the equation for the straight line response. The constant term is an approximate indication of the yield shear strain.

5.9.3 Limit pressure — Ghionna et al. (1990)

If, as in Fig. 5.15, the onset of "constant volume" behaviour can be identified, then the effective limit pressure, $p'_{\lim}$, can be evaluated. The concept is that the internal angle of friction is now at the critical state value and from this point onwards the area of material being sheared will remain the same. The following then applies:

$$p'_{\lim} = p'^{(cv)}_c \left[\frac{\varepsilon_v^{(cv)} + 1}{\varepsilon_v^{(cv)} - (\Delta A/A)} \right]^{\left(\frac{1-k_a^{cv}}{2} \right)} \qquad [5.41]$$

where

$$\frac{\Delta A}{A} = \varepsilon_c^2 - 2\varepsilon_c$$

The superscript (cv) signifies that these are the values for the parameters at the commencement of constant volume deformation. Equ. [5.41] uses the constant area ratio and this is a large strain construct. What evidence there is suggests that, given the same field curve, the results obtained from this approach compare well with the results for effective limit pressure derived from the modified Carter et al. solution.

5.9.4 Applying Manassero to cavity contraction — Whittle & Byrne (2020)

Because of insertion disturbance and the general noisiness of the loading curve with possibly several unload/reload cycles, there is an incentive to examine pressuremeter contraction data. However, the origin for the unloading event is uncertain—it is precisely defined for radial stress and hoop strain but the radial strain and hoop stress are highly dependent on the series of preceding calculations. Whittle & Byrne (2020) suggest this is only a problem if the purpose is to describe the complete unloading. If the intent is to obtain values for friction and the maximum shear stress, then it is sufficient to treat the turn-around coordinate as zero and then to apply the Manassero solution.

The modifications are straightforward. The point of maximum displacement (r_{mx}) and pressure (p'_{mx}) must be identified, and an unloading pressure (p'_{ul}) and unloading cavity strain (ε_{ul}) calculated:

$$p'_{ul} = p'_{mx} - p' \qquad [5.42]$$

$$\varepsilon_{ul} = -\frac{r_{mx} - r_c}{r_{mx}} \qquad\qquad [5.43]$$

where
  r_{mx} is the maximum radius when cavity contraction commences
  r_c is the current cavity radius

P'_{ul} and ε_{ul} replace p'_c and ε_t in (5.37). The resulting data are a mirror image of the true unloading path.

The effect of starting the contraction from a contrived origin is that the Manassero calculations prematurely run out of data with which to calculate the next step. This happens after the shear stress in contraction reaches a maximum and the friction angle reaches the constant volume state.

The true situation is sketched in Fig. 5.17 and an example of the procedure is given in Fig. 5.18. Points A and B indicate stress equality and maximum

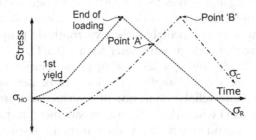

Figure 5.17 Radial and circumferential stress.

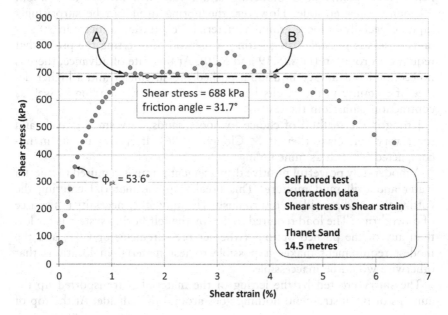

Figure 5.18 Shear stress vs shear strain.

circumferential stress. After point B, further calculation is inappropriate. As the authors emphasise, this is a strategy, not an analysis. The interesting characteristic which helps explain why curve fitting a drained test is difficult, is that the constant volume condition is reached at a small strain. It is then sustained at the constant volume condition for a much longer strain altera- tion. If it is assumed that this behaviour is a typical characteristic of a drained unloading with a purely frictional response then it is potentially a method for identifying the constant volume friction angle, ϕ'_{cv}. Initially, ϕ'_{cv} is a guess, and this is then modified until the plateau gives the same value.

5.10 TESTING LOOSE SAND

Loose sands are materials where the internal angle of friction is close to, or potentially less than, the critical state friction angle. Such soils may com- press under loading and makes them vulnerable to liquefaction. Obtaining high-quality samples from these soils is generally not possible. Of the forms of *in situ* testing that are available, the two routinely proposed are pres- suremeter testing and cone penetrometer testing (CPT).

Where possible, deploying a self-boring pressuremeter is the preferred option. Because the internal angle of friction can be derived without em- piricism, it is a more reliable indicator of liquefaction potential than the CPT (see Appendix D where the CPT and pressuremeter test are compared).

Self-boring is considered to be a slow form of testing, especially in comparison with a CPT. The loading of the ground and collection of suf- ficient data points at a moderate strain rate will always take about 30 minutes per episode. However, the boring itself can be surprisingly rapid, especially in loose granular material. Penetrating a metre in 90 sec- onds or less is possible using jetting, a method of inserting the probe that requires no rotation (Figs. 5.19 and 5.20). At this rate of advance, there is no economic advantage in using a pushed pressuremeter and a higher de- gree of certainty in the results because of access to the loading as well as contraction data from the test.

Jetting is the method of choice for loose sands, very small gravels and some very soft clays (Benoit & Clough, 1986). It is also usable in un- compacted fill such as mine tailings.

A moderately powerful positive displacement pump is required to inject a water and drill mud mixture. This breaks up the material entering the cutting shoe and provides a return flush. The only other necessity is a source of downthrust. The load required to deploy the self-boring system is far less than any of the pushed options (whether pressuremeter or CPT) and the modest requirement makes it possible to test material in locations that otherwise would be inaccessible.

The slurry created by the jetting of the material is transported up the annulus of the instrument; nothing goes around the outside. At the top of

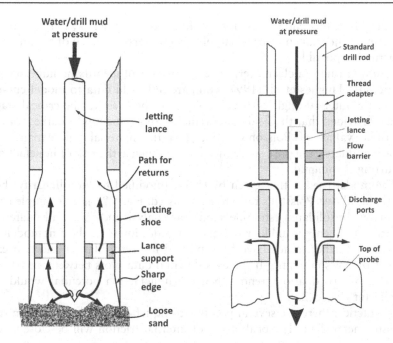

Figure 5.19 Jetting—lower end.

Figure 5.20 SBP jetting—top end.

the probe it is discharged into the open borehole and will return to the surface in a similar manner to standard drilling procedures. The variables are the rate of advance, the water flow rate and the distance between the tip of the jetting lance and the edge of the cutting shoe. The test that follows will indicate whether these have been set appropriately.

5.10.1 Testing for liquefaction susceptibility

Liquefaction refers to the uncontrolled effective stress reduction that can occur in some soils (usually young) due to the sudden application or cycling of a load. In loose sands, soft clays and uncompacted fills such as tailings the soil skeleton will want to contract, reducing the permeability. The load that would be supported by the soil skeleton is transferred to the pore water, and the pore water pressure rises. If that water cannot now permeate sufficiently quickly, then the effective stress is reduced and will keep falling whilst the load remains. Flow liquefaction (free-flowing material) occurs when the effective stress falls to near zero.

Earthquakes are more generally associated with cyclic softening, which can lead to the development of large shear strains (cyclic mobility) or cyclic liquefaction, both with potentially destructive consequences.

Testing the ground for liquefaction susceptibility or cyclic resistance can be done in the laboratory but it is precisely those soils that are difficult to sample where the tendency is greatest. In areas of acute risk, soil freezing is sometimes undertaken prior to sampling to preserve as much of the internal structure as possible.

In situ testing is a clear alternative as a means of identifying liquefaction propensity. Hughes et al. (1997) compare SBP field data to model curves where the internal angle of friction, ϕ'_{pk}, has been set to the critical state value. Provided that the field curve remains above the model curves for all plausible values of Poisson's ratio, then the material will expand on shearing (dilate), the pore space increases and hence the risk of liquefaction occurring is minimal.

The model used is that given by [5.35] (modified for non-linearity) because this allows Poisson's ratio to be adjusted. Fig. 5.21 is an example of a normally consolidated estuarine sand that analysis indicates will dilate on shearing and hence is unlikely to liquefy. In addition to the best-fit modelled data, two further trends have been plotted for zero dilation and two extremes of Poisson's ratio. If the experimental data plots between or below these two zero dilation trends, then the liquefaction potential would be significant (Fig. 5.22).

In practice, there are several pointers to liquefaction vulnerability in the pressuremeter data. The peak angle of internal friction will be close to or

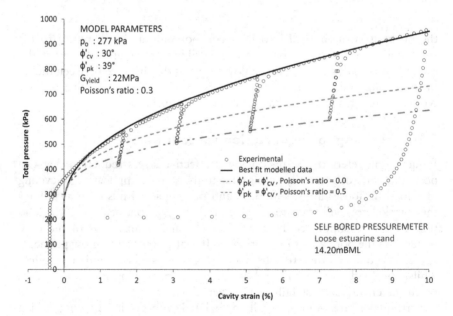

Figure 5.21 Sand that will dilate on shearing.

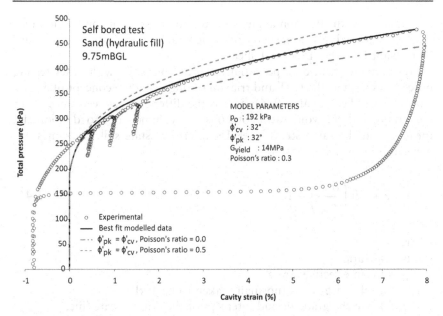

Figure 5.22 Sand with liquefaction potential.

less than the residual friction angle and the unload/reload cycles will give a similar stiffness/strain response because the mean effective stress is constant throughout the loading.

Using a pressuremeter to assess the *cyclic* susceptibility is an area of current research.

5.11 RECONCILING DATA — THE STATE PARAMETER AND RELATIVE DENSITY

Bolton (1986) shows that the internal angle of friction the angle at critical state, ϕ'_{cv}, and the angle of dilation, ψ, can be related by the following:

$$\phi'_{pk} = \phi'_{cv} + 0.8\psi \tag{5.44}$$

For dilation angles between 0 and 20° this arrangement gives, for all practical purposes, similar results to Rowe's dilatancy rule.

The importance of [5.44] is the representation of the mechanical behaviour of granular material shearing at angles above the critical line. Dilation is the process of particles overriding at points of contact, at stresses too low to permit crushing. ψ represents the additional volumetric strain that must be overcome before shearing at the critical angle takes place.

In this sense, the dilation angle is related to the state of compaction of a granular material. It is a property of plane strain deformation and may not be relevant for other modes of shearing.

Pressuremeter field data potentially can be compared with other testing procedures such as the CPT and triaxial testing using the concept of a state parameter, ξ (Been et al., 1986). This is the difference between the current void ratio and the void ratio for constant volume or zero dilation deformation at the same stress level. Using critical state soil mechanics nomenclature it is defined as:

$$\xi = e + \lambda \ln\left(\frac{\sigma'_{av}}{p_{ref}}\right) - \Gamma$$

[5.45]

where

e is void ratio

σ'_{av} is mean effective stress

p_{ref} is a reference stress (normally taken to be 1 kPa)

λ and Γ are the gradient and intercept of the critical state line.

Yu (1996) gives two expressions for deriving the initial state parameter from pressuremeter data. These apply the gradient of the response in log – log space, which may be taken from the loading phase (S in [5.5]) or contraction phase (S_{ul} in [5.16]):

Loading:

$$\xi_0 = 0.59 - 1.85\ S_\infty$$

[5.46]

$$\phi'_{pk} = 0.6 + 107.8\ S_\infty$$

[5.47]

Contraction:[2]

$$\xi_0 = 0.53 + 0.33\ S_{ul}$$

[5.48]

$$\phi'_{pk} = 6.6 - 18.4\ S_{ul}$$

[5.49]

In these arrangements ξ_0 is the initial soil state parameter, and all factors are empirical, derived from FEM, chamber tests and triaxial testing. The gradient of the loading response is written S_∞ to indicate this is the result for an infinitely long cavity. According to Yu, a length correction is needed for the finite length pressuremeter. A self-boring probe has a length to diameter ratio of approximately 6 and a gradient determined with this device must be reduced by a factor that depends on the stiffness of the material to give S_∞. Hence:

$$S_\infty = S\left[1.19 - 0.058 \ln\left(\frac{G_{yield}}{p'_o}\right)\right] \qquad [5.50]$$

No length adjustment is required for the unloading and one of the more interesting findings presented by Yu is confirmation that regardless of how a pressuremeter is inserted, all unloading responses are similar. However, only a low disturbance insertion method such as self-boring is expected to give representative loading data.

Negative values for ξ_0 indicate material that will dilate on shearing; positive values mean the material will compress.

The argument is attractive and it would be convenient if these relationships gave reasonable values. Unfortunately, comparing friction angles from fundamental analysis of pressuremeter data with those derived from state parameter correlations shows erratic agreement. In general, the state parameter values appear to over-estimate ϕ'_{pk}, in particular for the unloading case. A similar conclusion is reached by Wride et al. (2000), where state parameter values from CPT, pressuremeter and laboratory tests are compared.

This may be in part because deciding the slope of the unloading response is not always unambiguous.

This can be illustrated using an example employed by Yu (1996). Figs. 5.23 and 5.24 are the field curve and unloading analysis of a test in

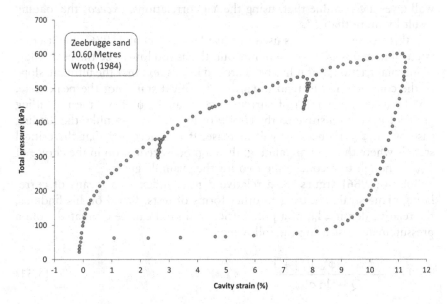

Figure 5.23 SBP test in sand (taken from Wroth, 1984).

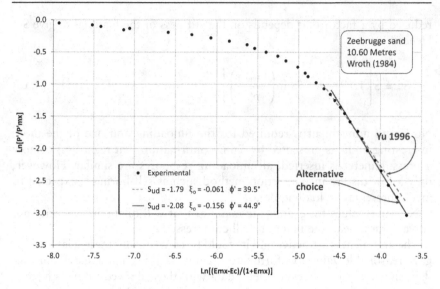

Figure 5.24 Unloading analysis (Yu, 1996)

sand taken from Wroth (1984). The parameters obtained are very sensitive to the choice of slope. That specified by Yu omits the data that follow reverse failure but gives a ϕ'_{pk} result similar to the loading (for a ϕ'_{cv} of 35°, Wroth[3] quotes 39°, Yu gives 39.5°).

However, an alternative and arguably more rigorous choice of slope that includes *all* the unloading data down to loss of contact with the borehole wall gives a ϕ'_{pk} value that, using the Yu correlations, *exceeds* the loading result by more than 5°.

If this steeper gradient is used to solve Houlsby et al. (1986), the friction angle obtained is 27°. As Yu points out, this is too low to be the peak angle of internal friction, but it has been argued in this text that the ultimate slope of the contraction will tend towards the critical state, not the peak state. The Houlsby calculation of current friction angle, ϕ', from the unloading gradient is very sensitive to the choice of input, ϕ'_{cv} — unlike the loading case, when ϕ'_{cv} reduces, ϕ' will increase. If it is accepted that the critical state is where the data are aiming, then a modest reduction in the choice of ϕ'_{cv} is enough to reconcile the two for the example given.

Bolton (1986) argues for a relative density index as a means of correlating plane strain shearing to other forms of tests. Based on his findings, the relative density, I_d, of a purely frictional sand can be estimated from a pressuremeter test using the following:

$$I_d = \frac{[(\phi'_{pk} - \phi'_{cv})/5] + 1}{Q - \ln \sigma'_{av}} \qquad [5.51]$$

Q is a shape factor relating critical state void ratio to mean effective stress and is approximately 10. σ'_{av} is the mean effective stress at yield and, using definitions already given, can be derived from the effective radial yield stress, p'_f:

$$\sigma'_{av} = \frac{p'_f}{3}\left(1 + \frac{1}{N} + \frac{1}{\sqrt{N}}\right) \qquad [5.52]$$

Applying this to the test shown in Fig. 5.6, I_d is 0.89, which is reasonable for a very dense sand.

All empirically derived relationships should be used with caution, but [5.51] seems to give plausible numbers. As there are strong indications that CPT tip load can be related to relative density, this is likely to be a more useful data comparison technique than the state parameter.

NOTES

1 The Manassero analysis does not follow the soil mechanics convention that compressive strains are positive. To avoid confusion with other analyses, this strain is denoted ε_t for tangential strain and is approximately $- \varepsilon_c$. When used to calculate the constant area ratio, the result will be negative for an increasing displacement.
2 S_{ul} is negative in this text so these correlations have a different sign to the versions given by Yu (1996).
3 Wroth (1984) states a dilation angle ψ of 9.5°, which must be a misprint. The parameters give $\psi = 5.5°$.

REFERENCES

Been, K., Crooks, J.H.A., Becker, D.E. & Jefferies, M.G. (1986) The cone penetration test in sands: part 1, state parameter interpretation. *Géotechnique* 36 (2), pp. 239–249.

Benoit, J. & Clough, G.W. (1986) Self-boring pressuremeter tests in soft clay. *Journal of Geotechnical Engineering* 112, pp. 60–78.

Bolton, M.D. (1986) The strength and dilatancy of sands. *Géotechnique* 36 (1), pp. 65–78.

Bolton, M.D., & Whittle, R.W. (1999) A non-linear elastic/perfectly plastic analysis for plane strain undrained expansion tests. *Géotechnique* 49 (1), pp. 133–141.

Carter, J.P., Booker, J.R. & Yeung, S.K. (1986) Cavity expansion in cohesive frictional soil. *Géotechnique* 36 (3), pp. 349–358.

Ghionna, V.N., Jamiolkowski, M. & Manassero, M. (1990) Limit pressure in expansion of cylindrical cavity in sand. *Pressuremeters. Third International symposium*, London: Thomas Telford, pp. 149–158.

Gibson, R.E., & Anderson, W.F. (1961) In situ measurement of soil properties with the pressuremeter. *Civil Engineering and Public Works Review* 56, 658, pp. 615–618.

Houlsby, G.T., Clarke, B.G. & Wroth, C.P. (1986) Analysis of the unloading of a pressuremeter test in sand, *The Pressuremeter and its Marine Applications: 2nd Int. Symp.*, pp. 245–262.

Hughes, J.M.O., Campanella, R.G. & Roy, D. (1997). A simple understanding of the liquefaction potential of sands from self-boring pressuremeter tests. *Proc. 14th International Conference* on *Soil Mechanics* and Foundation *Engineering* **1**, pp. 515–518.

Hughes, J.M.O., Wroth, C.P. & Windle, D. (1977) Pressuremeter tests in sands. *Géotechnique* **27** (4), pp. 455–477.

Janbu, N. (1963). Soil compressibility as determined by oedometer and triaxial tests. *Proc. 3rd Eur. Conf. Soil Mech.*, Wiesbaden, 2, pp. 19–24.

Manassero, M. (1989) Stress-strain relationships from drained self-boring pressuremeter tests in sands. *Géotechnique* **39** (2), pp. 293–307.

Rowe, P.W. (1962) The stress dilatancy relation for static equilibrium of an assembly of particles in *Contact. Proc. of the Royal Society* **269**, Series A, pp. 500–527.

Whittle, R.W. & Byrne, Y. (2020). Deriving stress/strain relationships from the contraction phase of pressuremeter tests in sands. Proc. Geovirtual 2020, Calgary, Canada, 14–16 Sept, Paper 107.

Withers, N.J., Howie, J., Hughes, J.M.O. & Robertson, P.K. (1989) Performance and analysis of cone pressuremeter tests in sands. *Géotechnique* **39** (3), pp. 433–454.

Wride, C.E., Robertson, P.K., Biggar, K.W., Campanella, R.G., Hofmann, B.A., Hughes, J.M.O., Küpper, A. & Woeller, D.J. (2000) Interpretation of in situ test results from the CANLEX sites. *Canadian Geotechnical Journal* **37**, pp. 505–529.

Wroth, C.P. (1984) The interpretation of in situ soil tests. 24th Rankine Lecture. *Géotechnique*, **34** (4), pp. 449–489.

Yu, H.S. (1996) Interpretation of pressuremeter unloading in sands. *Géotechnique* **46** (1), pp. 17–31.

Yu, H.S. & Houlsby, G. (1991) Finite cavity expansion in dilatant soils: loading analysis. *Géotechnique* **41** (2), pp. 173–183.

Chapter 6

Hydraulic Conductivity and Consolidation

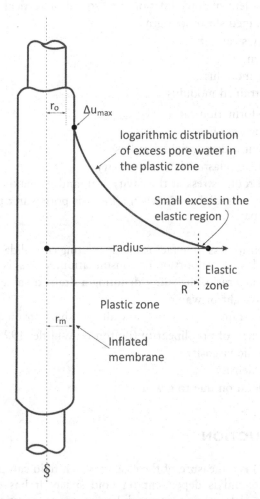

Figure 6.0 Radial distribution of excess pore water pressure around an expanding pres-suremeter in an undrained elasto-plastic soil (adapted from Clarke et al., 1979).

DOI: 10.1201/9781003200680-6

6.1 NOTATION

K_n Coefficient of permeability or hydraulic conductivity where n is h,v or m denoting horizontal, vertical or mean.

K_c Hydraulic conductivity at a length scale of 1

k Intrinsic permeability

U Excess pore water pressure

U_{mx} Maximum excess pore water pressure at cavity wall

u, u_{mx} Pore water pressure referenced to ground level and the maximum achieved

u_o Ambient pore water pressure

c, c_h, c_v Coefficient of consolidation, horizontal or vertical

c_u Undrained shear strength

ε_c Cavity strain, $(r - r_0)/r_0$

B Bulk modulus

G Shear modulus

M Constrained modulus

m Transformation ratio $\sqrt{(k_h/k_v)}$

n Scaling exponent

r, r_0 Current radius, original radius

R Radius of elastic/plastic boundary

p_c, p_{mx} Total radial stress at the cavity wall and its maximum value

p_u The total radial stress when all excess pore water pressure has dissipated

t Time

T_n Dimensionless time factor for calculating consolidation. Subscript n indicates proportion of consolidation, so T_{50} is 50%.

v' Poisson's ratio, the suffix denoting a drained value < 0.5

γ_w Unit weight of water

γ_c, γ_{ce} Shear strain at the borehole wall and shear strain at first yield

β Exponent of non-linearity (Bolton & Whittle, 1999)

η Dynamic viscosity

ρ Fluid density

g Acceleration due to gravity

6.2 INTRODUCTION

Permeability (k) is a measure of the ease at which fluid can pass through a porous medium and is dependent on void space. It has units of area. Techniques for determining permeability mean in practice finding the coefficient of permeability (K), termed hydraulic conductivity. This is a velocity and appropriate units are metres/second. The material must be in an *elastic* condition during the test because the measurement depends on the void space remaining unchanged.

Consolidation describes the process of reducing void space. This is a consequence of soil skeleton compression in the presence of a large hydraulic gradient caused by a significant *plastic* deformation. What is generally determined by experiment is the coefficient of consolidation (c), which has units of area and time, typically m²/year.

As both processes depend on the void space, the parameters are related and one can be used to derive the other:

Permeability k:

$$k = \frac{\eta}{\rho g} K \qquad [6.1]$$

where
K is hydraulic conductivity (m/s)
η is dynamic viscosity (Pa.s)
ρ is fluid density (kg/m³)
g is acceleration due to gravity (m/s²)

Given that for ground engineering purposes the fluid concerned is normally water, this becomes:

$$k = \frac{\eta}{\gamma_w} K \qquad [6.2]$$

where
γ_w is the unit weight of water

The coefficient of consolidation is obtained from hydraulic conductivity as follows:

$$c = \frac{K}{\gamma_w} M \qquad [6.3]$$

M is the coefficient of compressibility or the constrained modulus and can be derived from shear modulus:

$$M = 2G\left(\frac{1 - v'}{1 - 2v'}\right) \qquad [6.4]$$

The consolidation coefficient c and the hydraulic conductivity K both need a subscript to denote the direction of measurement. For a pressuremeter test in the horizontal plane, the parameters are c_h and K_h.

These are physical processes with a high degree of natural variability that can only be adequately quantified by experiment. Within a single borehole, it is not unusual to find 10 times the order of magnitude variation and close agreement between laboratory and field is not expected.

The dominant reason for this difference is sample size, but the direction of loading is also a significant factor. Cross-hole anisotropy is normal. At large scale, the manner of soil deposition means the ground is stratified, with a preference for water transportation in the horizontal plane. Soil deposits are frequently non-uniform at small scale due to partings of foreign material, fissures and joints. Proportionate replication of this diversity can be difficult in a small sample, but happens as a matter of course in the field setting.

The analyses for consolidation and permeability are straightforward. As [6.3] indicates, either measurement can be used indirectly to determine the other. This chapter differs from previous chapters in that the focus is primarily on technique, using a self-boring pressuremeter.

The concept of using pore pressure measurements taken during an SBP expansion test, as a means of obtaining consolidation parameters, was suggested by Hughes (1973). Clarke et al. (1979) give a procedure and solution that uses a modified SBP quick undrained test to determine the coefficient of lateral consolidation, c_h.

It is not usual to carry out self-boring for the sole purpose of determining consolidation or permeability. One advantage of the pressuremeter is that these parameters can be derived as a variation or add-on to a conventional test for strength and stiffness. In the case of consolidation the additional parameters are essential components of the consolidation solution.

There are self-boring probes dedicated to measuring permeability *in situ* (such as the PERMAC system manufactured by Roctest in Canada) but the use of these is dependent on making the economic case when the parameter concerned is so variable. The permeability testing described here does not use a purpose-built device. It is a standard Cambridge self-boring pressuremeter with some additional equipment that treats the SBP pocket and the connected drill string as a perfectly installed stand-pipe. This arrangement has been extensively researched and reported (Ratnam et al., 2005).

The advantage of the self-boring process for making these measurements is that a substantial volume of material is tested in as undisturbed a state as can be managed. Neither measurement is without compromise. In the case of consolidation testing, contraction data recorded after a consolidation phase are following a different stress/strain path to earlier data and will not assist in the interpretation of the overall test. Given the importance of the contraction data when modelling the test, this can be a significant loss.

Permeability testing in the way described means there is a simple method for altering the test geometry by a small retraction of the self boring probe. This greatly expands the possibilities of the test and interpretation but the compromise is possible smearing and some stress relief as the permeated pocket is unsupported except by water.

The only type of pressuremeter being considered for either test is one placed by self-boring because only these devices are fitted with piezometers. In principle, any pressuremeter test could be used to determine consolidation if one were to assume the decay of total stress in a constant strain test were equivalent to the decay of excess pore water pressure. However, the materials where a pre-bored test might be carried out are precisely those where this assumption is unsafe.

The tests for permeability and consolidation are applicable to soil-like material only. It is possible that a completely weathered rock could be tested in similar ways but to date there are no data where this has been attempted. In more competent rock, permeability is likely to be dominated by flow through discontinuities.

6.3 CONSOLIDATION TESTING

The analysis for the pressuremeter consolidation test arises from considering the problem of the driven pile. If a pile is driven into a cohesive soil of low permeability then the material adjacent to the pile is taken to limit pressure and there is a consequent build up of excess pore water pressure at the interface. Over time, the hydraulic gradient that has been created forces the excess water pressure through the soil matrix and the material consolidates. Soil particles move back towards the pile; there is an increase in the effective stress and an associated increase in strength.

Consolidation around a driven pile or a cone penetrometer (CPT) is a constant strain process because the dimensions of the object define the cavity radius. A pressuremeter cavity expansion test used for consolidation purposes is analogous to this—the difference is only the degree of strain imposed on the material. Because the pile (or CPT) is driven into virgin ground, the shear strain is 100%. For the self-boring probe, the shear strain will typically be less than 20%.

Randolph and Wroth (1979) is a closed form solution for the consolidation around a driven pile assuming radial pore water flow, and an elastically deforming soil skeleton under plane strain conditions. The solution provides time factors, T, relating proportion of consolidation to the ratio U_{mx}/c_u where U is excess pore water pressure and c_u is undrained shear strength. This is the natural log of the ratio of the plastically deformed radius to elastic radius (Fig. 6.0).

The magnitude of the excess pore water pressure and the duration of this process is calculable from the material properties for stiffness, strength and permeability. The permeability is unknown, so a consolidation test monitors the dissipation process for a period of time until a fixed proportion of the excess has dissipated, typically 50%.

The Randolph and Wroth time factor T50 is used by Clarke et al. (1979) to analyse the case of a quick undrained pressuremeter test where the expansion is stopped and 50% of the generated excess pore water allowed to dissipate. This solution can be written as:

$$c_h = \frac{T_{50} r_m^2}{t_{50}}$$

[6.5]

where

c_h is the coefficient of lateral consolidation

T_{50} is the dimensionless time factor for 50% consolidation

r_m is the expanded radius of the pressuremeter in metres

t_{50} is the observed or calculated time for 50% of the excess pore water pressure to dissipate.

6.3.1 Constant strain consolidation test example

Fig. 6.1 is a quick undrained test is carried out in a saturated clay to a significant expansion. At about 8.2% cavity strain, the expansion has been stopped and the diameter of the test cavity has been maintained at this value. Excess pore water pressure generated by the expansion is allowed to dissipate (Fig. 6.2). These data come from a self-bored test carried out at the UK soft clay test site, Bothkennar in Scotland, reported by Ratnam et al. (2002).

Maintaining a constant cavity diameter with the expanded self boring probe is difficult to achieve manually. A semi-automatic system monitors the tendency of the borehole to expand as the excess pore pressure decreases and compensates by adjusting the internal pressure. The diameter of the cavity stays fixed (Fig. 6.3). One consequence is that in addition to the

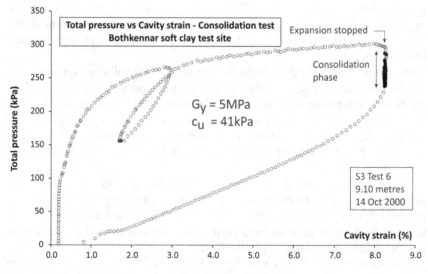

Figure 6.1 SBP Consolidation test in soft clay.

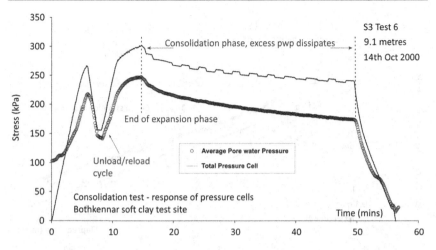

Figure 6.2 Pressure cell response.

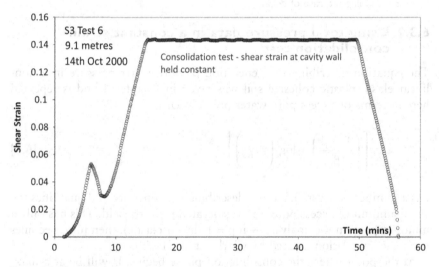

Figure 6.3 Constant strain.

smooth pore pressure decay, there is a total pressure decay trend that in the example looks very similar. The "bumpiness" is the control system slightly overshooting as it hunts for the optimum value of internal pressure.

The easiest and most reliable way of obtaining t_{50} is to wait the length of time required for 50% of the excess to dissipate (Fig. 6.2). Sometimes this is not possible. Fig. 6.4 plots the normalised decay (U/U_{mx}) trends on semi-log scales and finds the slope. This gives a reasonable fit to the data for all but the first minute of dissipation. Given the slope, the time for any proportion of dissipation can be calculated. In the example, the differences between observed time and calculated time are insignificant.

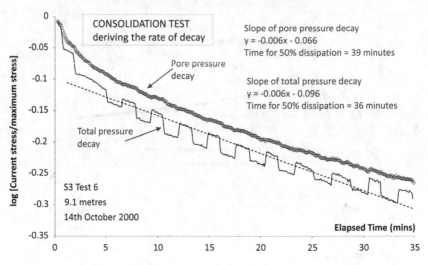

Figure 6.4 Finding the rate of decay.

6.3.2 Using total pressure data in a constant strain consolidation test

The equation describing the generation of pore water pressure in a non-linear elasto-plastic cohesive soil was given in Chapter 4 and is repeated here in terms of excess pore water pressure U:

$$U = c_u \left[\left(\frac{1 - \beta}{\beta} \right) + \log_e \left(\gamma_c / \gamma_{ce} \right) \right] \qquad [6.6]$$

β is a number between 0.5 and 1 describing non-linearity and quantifies the small amount of excess pore water generated prior to yield. This has only a minor impact on the analysis — if $\beta = 1$ (linear elastic), then [6.6] becomes the same calculation as used by Clarke et al. (1979).

At the point where the consolidation phase begins, U will be at a maximum, U_{mx}. The hydraulic gradient created by the cavity expansion forces U towards zero.

Similarly, the total pressure will have a maximum value, p_{mx}, when the consolidation phases commences. If it is assumed that the decay of total pressure matches that of the pore water pressure, then this occurs when the total stress has reduced to the value, p_u, defined as:

$$p_u = p_{mx} - u_{mx} \qquad [6.7]$$

Once the consolidation phase is initiated, p_u is deducted from all total pressure readings to get the decay trend that represents the dissipation of excess pore water.

If the total stress and pore pressure decay at the same rate then at the cavity wall the effective radial stress is unchanged. This seems a contradiction. As consolidation is tending to increase the available strength the expectation is that the radial stress will fall at a lesser rate. In practice, this is not always evident.

This is addressed to some extent by Randolph and Wroth. It is suggested that the non-linearity of the soil influences the magnitude of stress change but not the rate of dissipation. This description is partially supported by FEM experiment.

Clarke (1990) recommends using total pressure decay instead of pore pressure decay as being the more reliable measure on practical rather than theoretical grounds. This is not good practice. There are certainly data (such as the example given in Fig. 6.4), indicating that it can be a close match to the pore pressure trend, but this is not invariably the case. Fig. 6.5 is taken from a test in Gault Clay. The pore pressure data give a consolidation coefficient of 10.7 m^2/yr and a hydraulic conductivity rate of 1.2 × 10^{-10} m/s, both of which are reasonable results in this heavily over-consolidated clay. The normalised total stress trend departs from the normalised pore pressure trend after about 10 minutes and, if used, would lead to unrealistically low values.

The obvious explanation for the divergence is that the material is consolidating and the effective radial stress is increasing. Why this should be so quickly apparent in this heavily over-consolidated material and not in the lightly over-consolidated clay is unexplained. It will be related to the propensity for the Gault Clay to dilate on shearing and the

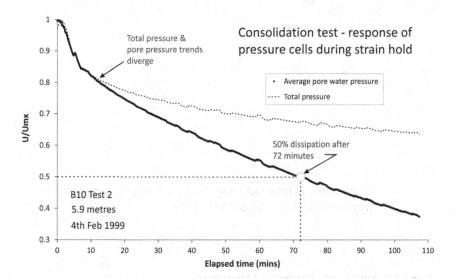

Figure 6.5 Consolidation test in Gault Clay.

Bothkennar clay to compress. Based on this, it seems reasonable to predict that monitoring the total stress decay may only give a plausible time measurement in compressive clays. This is developed further in section 6.4.4.

6.3.3 Time factors

Fig. 6.6 is a representation of the consolidation curves produced by Randolph and Wroth. This version has been created by fitting the published response with polynomial functions to allow software to incorporate them. The factor "T" relates consolidation, the radius of the expanded area and time:

$$T = \frac{ct}{r_o^2}$$
[6.8]

where
  c is consolidation
  t is time
  r_o is the original radius

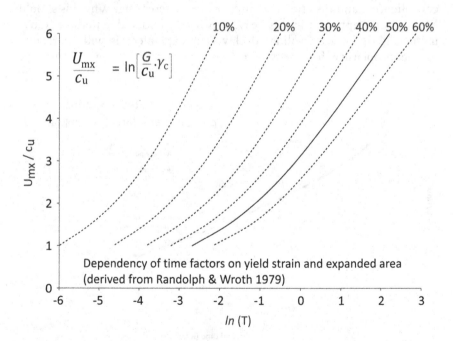

Figure 6.6 Consolidation time factors (dimensionless).

All the units cancel, leaving T as a dimension-less value. For the T_{50} trend, the polynomial used to find T is:

$$T_{50} = Exp[ax^4 + bx^3 + cx^2 + dx + e] \qquad [6.9]$$

where
   a = −0.0135 b = 0.197 c = −1.133 d = 3.8 e = −5.5504

   $x = U_{mx}/c_u$

Equ. [6.9] reproduces the published trend with a correlation coefficient better than 0.999. Similar functions can be produced for the consolidation curves at different proportions.

6.4 CALCULATING CONSOLIDATION—WORKED EXAMPLE

Assuming that both the total stress and excess pore pressure data are available as decay trends, the interpretation can be done in two ways. The simplest is to use the pore pressure decay and the measured time, as this requires the fewer assumptions. Alternatively, the total pressure decay trend if deemed representative, can be used with missing information derived from closed-form elasto-plastic solutions.

6.4.1 Using the pore pressure decay

Generally, when using the self-boring probe, the pore pressure transducers are referenced to ground level conditions. The measurement they make is total pore water pressure, u. The interpretation uses *excess* pore water pressure, U. Assuming this is the case, the following data are needed:

- Ambient pore water pressure, u_o (normally calculated from the water table).
- Maximum excess pore water pressure just prior to commencing the consolidation phase, u_{mx}.
- Radius of the cavity at the end of the expansion phase—this is the measured displacement added to the at-rest radius of the probe, giving r_m.
- The time taken for 50% of the *excess* pore water pressure (U_{mx}) to dissipate.
- A value for undrained shear strength, c_u, from analysing the test data prior to the consolidation phase commencing.

The steps are these:

- Derive the maximum excess pore pressure, $U_{mx} = u_{mx} - u_o$.
- Calculate U_{mx}/c_u.

- Use this is to find the time factor T50 from [6.9].
- Equation [6.5] can now be solved and will give an answer in units of m^2/min. Following convention, the lateral consolidation is generally expressed in units of m^2/year so the result of the calculation is multiplied by 525600.

For the test shown in Fig. 6.1:

Table 6.1 Measured dissipation

r_m (m)	c_u (kPa)	U_{mx} (kPa)	U_{mx}/c_u	T50	t_{50} PPC (mins)	c_h (m^2/yr)	K_h (m/sec)
0.0446	41	163	3.97	1.915	39	52	1.2×10^{-9}

6.4.2 Using the total pressure decay

As an exercise, it is assumed that no data are available for the pore water pressures. The following input data are required:

- A full analysis of the test data recorded prior to the consolidation phase commencing
- Radius of the cavity at the end of the expansion phase, r_m.
- The time taken for 50% of the *excess* total pressure to dissipate. The starting point will be p_{mx}, the ultimate end point will be $p_{mx} - U_{mx}$. Note, U_{mx} is obtained from [6.6].

Using [6.6], the ratio U_{mx}/c_u is calculated. This is then used to find the time factor T50 from [6.9].

Finally, the observed time for 50% decay is taken from Fig. 6.2 or Fig. 6.4. For the test shown, using calculated data for U_{mx} and t_{50} from total pressure decay:

Table 6.2 Calculated dissipation

r_m (m)	c_u (kPa)	U_{mx} (kPa)	U_{mx}/c_u	T50	t_{50} (mins)	c_h (m^2/yr)	K_h (m/sec)
0.0446	41	133	3.24	1.086	36	32.3	7.5×10^{-10}

6.4.3 Variation in the results

Within experimental limits, the two approaches to obtain c_h give broadly similar results. At the same site and depth, using a linear elastic/perfectly plastic solution, Clarke (1990) reports 64 m^2/year using total pressure but with a different definition of where the decay trend is aiming.

It is possible to mix aspects of the measured approach and theoretical approach to achieve a closer match, but that is not the purpose of the demonstration. The major difference between the two is the calculation of

U_{mx} compared to the measurement of U_{mx}. Because the measured value is higher than the predicted value, it is the preferred choice—the greatest risk in consolidation testing is the premature commencement of dissipation before the expansion phase is completed but that is not the case here. The difference is indicating the limits of the analysis for strength and stiffness. This material is not perfectly plastic—it has a marked peak and residual strength, so pore water generation is driven by the peak strength, whereas the predicted trend is biased towards the average strength.

As an example of the effect of partial dissipation prior to the consolidation phase, consider the pwp trend shown in Fig. 6.4 and imagine it starting from the 15 minutes elapsed time point. U_{mx} will now be 68% of the true value and 84% of the calculated value (see below).

Table 6.3 Calculations from premature loss of pore water pressure

r_m (m)	c_u (kPa)	U_{mx} (kPa)	U_{mx} meas. (kPa)	U_{mx}/c_u	T50	t_{50} (mins)	c_h (m²/yr)	K_h (m/sec)
0.0446	41	133	112	3.24	1.086	45	25.6	5.9×10^{-10}

Given the data in Table 6.3, the interpretation recognises that the new "measured" U_{mx} is significantly less than the calculated value and hence prefers the calculated. The t_{50} value is still 50% of the new maximum but, because the dissipation is now not observed for long enough, it will have to be derived from the slope of the trend.

The calculations for c_h apply the theoretical U_{mx} for the purposes of finding T50. It is the best that can be done with imperfect data, but the result is half that derived from the analysis of the complete dissipation trend. Tests showing some drainage prior to initiating the consolidation phase always under-estimate the consolidation coefficient.

6.4.4 Other dissipation rates

The 50% or half-life interpretation of the consolidation test is standard and it is the method followed by piezocone dissipation tests. There is nothing fundamental in the choice of 50% and other rates can be applied using the Randolph and Wroth solution. Fig. 6.7 is an example from a self-bored test in a lightly over-consolidated lancustrine clay found in Switzerland. Details of the site and the research programme from which this example has been taken are given in Schneider et al. (2022).

The dissipation trend was observed for longer than normal and consolidation values have been derived from 40%, 50% and 60% dissipation proportions. The results are given in Table 6.4.

The U_{mx}/c_u ratio of 2.92 for this material has been used to find the time factor for the relevant dissipation rate. Fig. 6.7 has been used to find the elapsed time for the particular percentage of dissipation.

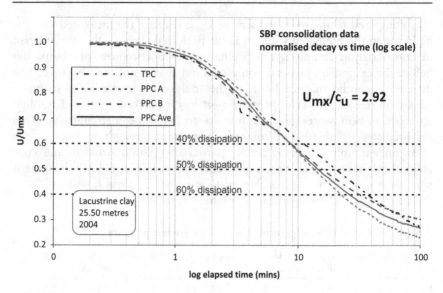

Figure 6.7 Long duration dissipation test.

Table 6.4 Results table for Fig 6.7

Decay (%)	Time factor	PPC mins	c_h (PPC) (m^2/yr)	TPC mins	c_h (TPC) (m^2/yr)
40	0.469	8.8	54.4	11.2	42.7
50	0.826	14.6	57.7	37.2	22.7
60	1.363	26.4	52.7	76.4	18.2

For the mean pore pressure decay, there is very good agreement for all three dissipation levels. For the total pressure decay the agreement is poor, especially for higher proportions of dissipation. This is another example of the total pressure response being controlled by a local increase in material strength.

The results and the plot suggest that a smaller dissipation percentage gives better agreement between the two decay trends. Based on Figs. 6.5 and 6.7, a dissipation proportion of 20% would give total pressure decay results indistinguishable from the pore pressure decay. If this were demonstrated to be the general case, then instruments such as pre-bored probes without piezometers would be able to measure consolidation. This is an interesting area and requires more investigation.

As things are, the evidence is mixed. Greater dissipation proportions are preferred because the starting condition has a marked effect on the initial decay gradient. A decay of 50% therefore seems a reasonable compromise between quality and cost, but requires direct measurement of the pore pressure response.

6.4.5 Permeability coefficient from consolidation

If a full analysis of the test data prior to the consolidation phase has been carried out, then G_y the shear modulus at yield can be calculated from the shear strength, c_u, and the shear strain at first yield, γ_{ce}:

$$G_y = \frac{c_u}{\gamma_{ce}} \qquad\qquad [6.10]$$

G_y is used to derive hydraulic conductivity from consolidation. As indicated by [6.4] a value for drained Poisson's ratio, ν, is also required and in the absence of better information it is customary to assume 0.2. The connection between consolidation and the coefficient of hydraulic conductivity is given by [6.3], but in practice what is used is:

$$K_h = \frac{c_h}{G_y} K_{Const} \qquad\qquad [6.11]$$

where
    K_h is lateral hydraulic conductivity in m/s
    c_h is coefficient of horizontal consolidation in m^2/year
    G_y is the shear modulus at yield strain (in MPa)

$$K_{Const} = \frac{9.81\,(1 - 2v')}{1000 \times 3600 \times 24 \times 365 \times 2(1 - v')} = 1.167 \times 10^{-10}$$

Equ. [6.11] has been used to calculate the K_h values in Tables 6.2 and 6.3 (Fig. 6.8).

6.5 PERMEABILITY TESTING

Direct measurement of the hydraulic conductivity can be done using a standard self-boring pressuremeter. The "permeameter" consists of a self-bored pocket, the SBP itself just being part of the water delivery pipe. Even for an SBP arranged to drill slightly oversize the first 0.25 metres of the probe is always an exact fit to the ground and sees the external stress. This provides a natural seal.

Once in position in the ground, the drill string is closed off. Water is pumped down the string, and the only available exit is out of the cutting shoe on the foot of the instrument and into the strata. The system is intended for low permeability soils and so the water is delivered at a constant rate of flow (as suggested by Harwood et al., 1995). For a given flow, there is an

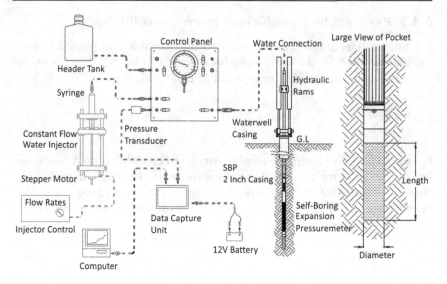

Figure 6.8 Schematic of self-boring pressuremeter used as permeameter.

equilibrium pressure that is related to the permeability and the geometry of the arrangement of parts.

If the flow is now increased, a new and higher equilibrium pressure will be found. A test is a series of these steps (Fig. 6.9) and the required output is a set of coordinates of flow rate against steady-state pressure (Fig. 6.10).

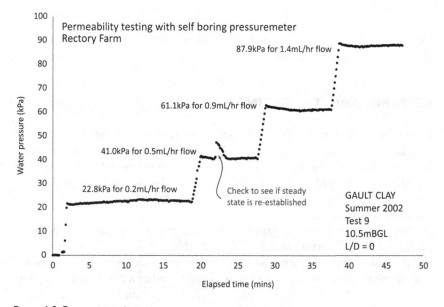

Figure 6.9 Pressure vs time.

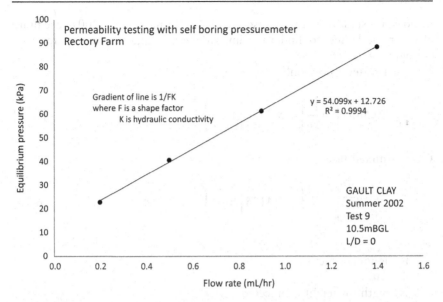

Figure 6.10 Pressure vs flow.

The geometry is unusual in that the exposed surface at the cutting shoe edge is always included. This method determines K_m, the mean hydraulic conductivity. However, if the probe is retracted a little then the height of the injected cavity is increased and a test repeated. The greater the height, the more the lateral permeability comes to dominate the measurement. A series of tests with different heights potentially gives the relationship between the vertical and horizontal permeability.

6.6 THE PERMEABILITY ANALYSIS

The solution for a permeameter test uses an equation of the following form derived from Darcy's flow rule:

$$K = Q/FH \hspace{4cm} [6.12]$$

where
   K is hydraulic conductivity
   Q_∞ is flow under steady-state conditions
   H is the applied head of water
   F is a shape factor depending on the geometry of the permeameter setup and variations in the horizontal to vertical permeability.

Hvorslev (1951) gives a series of solutions for the shape or intake factors in [6.12], depending on the configuration of the injected pocket. These solutions are approximations and should not be used with the short height of pocket tested by methods described here. Ratnam et al. (2001) use finite element modelling to find the dimensionless intake factor F/D for two configurations. These are:

Case 1 (lateral flow only)

$$\frac{F}{D} = 0.5691 \left[m\frac{L}{D} \right] + 5.2428 \left[m\frac{L}{D} \right]^{0.5} \qquad [6.13]$$

Case 2 (mixed flow)

$$\frac{F}{D} = 1.1872 \left[m\frac{L}{D} \right] + 2.4135 \left[m\frac{L}{D} \right]^{0.5} + 3.1146 \qquad [6.14]$$

where
   F is the intake factor
   D is diameter
   L is length or height of injected cavity
   m is the transformation ratio for anisotropic soils $\sqrt{[k_h/k_v]}$

Fig. 6.11 sketches the two configurations examined. Case 1 applies to the majority of specialised permeameters. Changing the L/D ratio (without removing the probe from the ground) is not an option for the purpose-built devices.

Fig. 6.12 is taken from the Bothkennar test site and shows the effect of modifying the geometry from a sequence of four tests, each of greater height than the previous (the results are given in Table 6.1). What is measured is K_m. The values for the horizontal and lateral permeability coefficients are derived, assuming that the anisotropy ratio $K_h/K_v = 2$ as this gives the best agreement between all values. The long L/D ratio was used to produce a result that could be compared with measurements already made at this site using a fixed geometry dedicated permeameter. Note that the lateral conductivity result inferred from a consolidation test at 9 metres (given in Table 6.1) is 1.2×10^{-9} m/s, less than the L/D 4 value for K_h of 8.17×10^{-9} but in the context of this complex measurement is encouraging agreement (Table 6.5).

The calculation for the results in this table is:

$$K_m = g/(F \times \text{Slope} \times 3600 \times 10^{-6}) \qquad [6.15]$$

where
   g is acceleration due to gravity
   F is calculated from [6.14] and is given dimensions of metres by multiplying by the cavity diameter

Slope is taken from a plot such as Fig. 6.10 and is pressure in kPa divided by flow in mL/hour

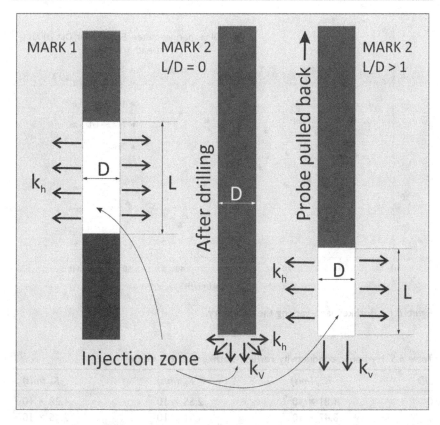

Figure 6.11 The two permeameter configurations.

This gives a result for K_m in units of m/sec. Given K_m:

$$K_h = mK_m$$
$$K_v = K_m/m$$

[6.16]

where
m is the transformation ratio used in [6.14]

6.6.1 Scale

The ability to vary the geometry reveals a little reported characteristic of permeability measurements, which is the scale or aperture effect. As the surface area of the permeated material increases, any heterogenous characteristics become disproportionately significant and affect the measurement. Ratnam et al. (2005) consider this phenomenon and propose the following exponential function relating hydraulic conductivity to scale of measurement:

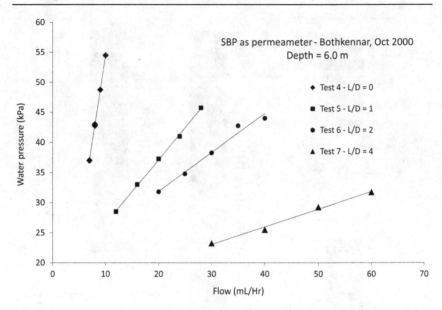

Figure 6.12 Bothkennar—varying the geometry.

Table 6.5 Hydraulic conductivity results, Bothkennar, 6.0 metres

L/D	K_m (m/s)	K_h (m/s)	K_v (m/s)
0	1.81×10^{-9}	2.55×10^{-9}	1.28×10^{-9}
1	3.47×10^{-9}	4.91×10^{-9}	2.45×10^{-9}
2	4.00×10^{-9}	5.65×10^{-9}	2.83×10^{-9}
4	5.77×10^{-9}	8.17×10^{-9}	4.08×10^{-9}

$$K = K_c (\text{length scale})^n \qquad [6.17]$$

where
K_c is the hydraulic conductivity value at a length scale of 1
n is the scaling exponent.

In a homogeneous material, the exponent n will tend to zero. The Bothkennar soft clay results indicate an exponent of 0.8.

"Length scale" can be any associated variable measure—Ratnam et al. (2005) use the volume of injected fluid.

6.6.2 Potential issues with the testing method

The primary issues with this style of test are:

- Temperature susceptibility
- De-airing
- Leakage
- Smearing

The permeameter adaptation means adding components to the self-boring pressuremeter system that exists at ground level. The flows involved are often tiny and temperature variation (because it affects the injected volume) can be a problem. The exposed length of the drill string is arranged to be as close to ground level as possible. Temperature data are collected as part of the data stream, not so much as a means of correction but of assessing the validity of the measurements (Fig. 6.13).

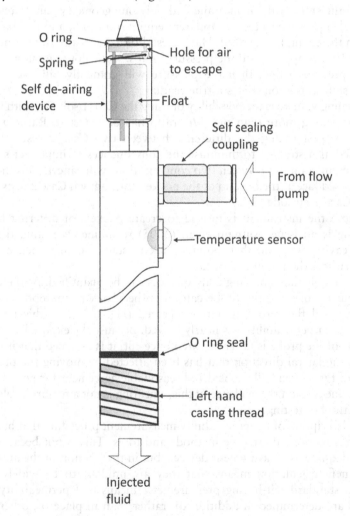

Figure 6.13 Top end seal with de-airing facility.

De-airing the drill string is essential and is done by a combination of procedural routines and use of an automatic venting device at the top of the casing column. The drill string itself uses double O-ring seals for integrity. Prior to the equipment being deployed, the probe has the cutting shoe blocked off and a pseudo test with the full control gear is run to prove that it is water-tight.

In the ground itself, the presence or absence of leakage paths is established by using the membrane of the probe. Given a test such as Fig. 6.9, after finding the first steady state, the membrane of the probe is slightly inflated. If there had been leakage around the probe then this would shorten the height of the test cavity and the pressure would rise without any change to the flow rate. If it does, then the height of the cavity is from the cutting shoe to the start of the membrane and only one geometry can be tested.

It is also possible to test an apparent equilibrium step, and this has been done in the example (Fig. 6.9). Having established a steady state, a pulse of greater flow is applied and the pressure jumps. If the flow rate now returns to the previous value, then the pressure will gradually fall back to the steady-state condition if it is a true reading.

Intriguingly, there is the possibility of using the Case 1 solution with Case 2 data, assuming more than one L/D ratio has been tested. Ratnam et al. (2005) suggest taking the difference between two Case 2 tests. This is proposed as a strategy to minimise the consequences of imperfect sealing. Anything in common between two configurations will cancel. This includes the exposed face at the bottom of the pocket, turning two Case 2 tests into a single Case 1 episode.

Because the instrument is moved to create pockets of differing height, smearing is an issue. Ratnam et al. (2005) recommends testing the zero length cavity first, and then pulling back in stages so that each exposed length suffers the minimum smear.

It is possible that smearing consequences can be quantified. At the end of the sequence of tests, the probe can be pushed back to the bottom of the borehole and the zero length test repeated (Fig. 6.14). In this example, the apparent permeability has nearly halved. Because the exposed surface at the base of the probe is never smeared once cut, it is assumed that it is only flow in the lateral direction that has been affected by moving the probe. A factor of two is generally considered reasonable agreement for permeability measurement so on the evidence of this, smearing is not a major problem for this method of testing.

The description of the permeability measurement procedure that has been given has ignored alternative methods and tools. This is not because they are inadequate as measurement devices, but in the opinion of the authors a cost/benefit calculation means that they are unlikely to be widely used. Using a standard self-boring pressuremeter means that permeability coefficients are determined in addition to, rather than in place of, other measurements for soil properties. The method is intended for fine-grained soils

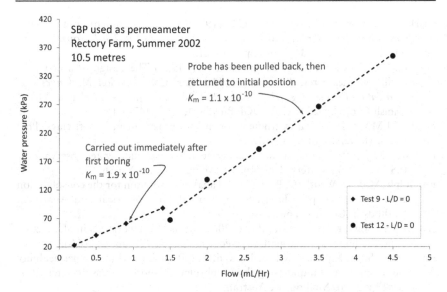

Figure 6.14 Investigating smearing.

with hydraulic conductivity values below 10^{-7} m/s. However, on one occasion, when the pressuremeter was inadvertently installed in a more permeable layer that exceeded the flow rate capability of the system, the drill string allowed a falling head test to be carried out. It is an unusually versatile arrangement.

6.6.3 Permeability or consolidation?

It was noted previously that the Bothkennar results for hydraulic conductivity derived from consolidation tests were lower but of similar magnitude to those values for hydraulic conductivity derived from direct measurement. Fig. 6.5 is a test in heavily over-consolidated Gault Clay. This example gives a K_h of 1.2×10^{-10}. The nearest available direct measurement for K_h reported by Ratnam et al. (2005) for this location is 2.25×10^{-10}.

In general, hydraulic conductivity results reported from SBP consolidation tests do seem to be slightly lower than SBP direct measurement results but broadly similar. The reverse situation applies regarding consolidation.

REFERENCES

Bolton, M.D. & Whittle, R.W. (1999) A non-linear elastic/perfectly plastic analysis for plane strain undrained expansion tests. *Géotechnique* **49** (1), pp. 133–141.

Clarke, B.G. (1990) Consolidation characteristics of clays from self-boring pressuremeter tests. *Proc. 24th Annual Conference on Engineering, Group of the Geological Soc.: Field Testing in Engineering Geology*, Sunderland, pp. 19–35.

Clarke, B.G., Carter, J.P. & Wroth, C.P. (1979) In situ determination of the consolidation characteristics of saturated clay. *Proc. of the 7th European Conf. on Soil Mechanics*, Brighton, pp. 202–213.

Harwood, A.H., Clarke, B.G. & Araruna, J.T. (1995) The design of a new self-boring permeameter, *Proc. of the 11th Eur. Conf. on Soil Mechanics and Foundation Engineering (XI ECSMFE '95)*, 28 May-1 June, Copenhagen, Vol. 1, Danish Geotechnical Society, DGF-Bulletin 1, pp. 1.147–1.154.

Hughes, J.M.O. (1973) An instrument for in situ measurement in soft clays, PhD thesis, University of Cambridge.

Hvorslev, M.J. (1951) Time-lag and soil permeability in ground water observations. U.S. Army Eng Waterw Exp Stn, Vicksburg, Bulletin No 36.

Randolph, M. F. & Wroth, C. P. (1979) An analytical solution for the consolidation around a driven pile. International Journal for Numerical and Analytical Methods in Geomechanics3, pp. 217–229.

Ratnam, S., Soga, K. & Whittle, R.W. (2001) Revisiting Hvorslev's intake factors using the finite element method, *Géotechnique* 51 (7), pp. 641–646.

Ratnam, S., Soga, K., Mair, R.J. & Bidwell, T. (2000) A novel in situ permeability measurement technique using the Cambridge self-boring pressuremeter, *Proc GeoEng 2000*, Melbourne, Australia.

Ratnam, S., Soga, K. & Whittle, R.W. (2002) Self-boring pressuremeter permeability measurements in Bothkennar clay. *Géotechnique* 52 (1), pp. 55–60.

Ratnam, S., Soga, K. and Whittle, R.W. (2005) A field permeability measurement technique using a conventional self-boring pressuremeter. *Géotechnique* 55 (7), pp. 527–537.

Schneider, M., Whittle, R.W. & Springman, S.M. (2022) Measuring strength and consolidation properties in lacustrine clay using piezocone and self-boring pressuremeter tests. *Paper accepted for publication by Canadian Geotechnical Journal.*

Chapter 7

Interpretation of Tests in Rock

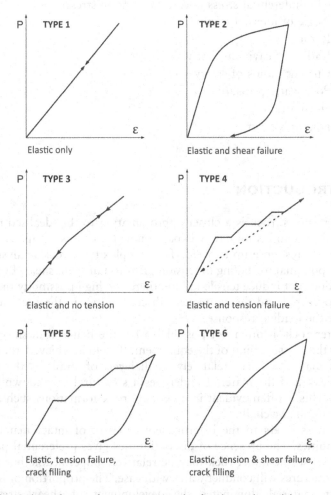

Fig 7.0 The six basic forms of pressure:strain plot for the rock model.

DOI: 10.1201/9781003200680-7

7.1 NOTATION

ε	Cavity strain
c_u	Undrained shear strength
E	Elastic (Young's) modulus
G	Shear modulus
P	Total pressure measured at the cavity wall (radial stress)
p_o	Initial cavity reference stress (total)
p'_o	Initial cavity reference stress (effective)
P_{mx}	Maximum pressure at the end of a cavity expansion
σ_r	Radial stress
σ_θ	Circumferential stress (also termed hoop stress)
ϕ	Angle of internal friction
r	Radius
r_f	Radius of cavity at first yield
r_c	Current radius of cavity
U	Pore water pressure
γ	Shear strain
ν	Poisson's ratio

7.2 INTRODUCTION

In some respects, rock is a closer approximation to the idealised material assumed by solutions for cavity expansion as it exhibits an approximately linear elastic response up to yield. The complexity of rock analysis arises from the potential for failing in tension prior to failure in shear. Even if the loading does not induce tensile failure, there is a high possibility that it has already occurred and the consequent discontinuities in the rock mass will influence the loading response.

In better rock, it often not possible to fail the material in shear before reaching the pressure limit of the equipment. If yield is achieved, the loading response may plateau at relatively small levels of strain as the internal bonds break and the material undergoes a structural breakdown or pore collapse. This is often evident in carbonate rock formations such as limestone, dolomite or chalk.

Rock mass refers to the heterogeneous matrix of intact material and discontinuities. The presence of discontinuities gives preferential paths for water flow, so the intact rock may be relatively impermeable but the network of fractures will conduct water with ease. The proportion of fractures to intact material dominates the development of shear stress. The assumptions of homogeneity and isotropic distribution of properties that are intrinsic to cavity expansion solutions will generally be inappropriate. The extent to which this invalidates the use of such solutions when applied to a rock test is a matter of judgement. The purpose of this chapter is to inform that judgement.

Pressuremeter tests in rock are two types:

1. Tests in weak to very weak rock, evidenced by substantial plastic development
2. Tests in weak to moderately strong rock, where all observed response is elastic

One of the virtues of high-resolution pressuremeters is that *relative* displacements of less than one micrometre can be measured reliably. It is possible to interpret with confidence small details of loading paths where there are minimal indications of a plastic response.

The interpretation commences by identifying what type of rock behaviour is being seen. It is a visual rather than arithmetic process and to aid in this, a simple model of potential rock behaviour is described and developed. This is illustrated with some examples from field tests.

7.3 POCKET FORMATION

For the most part, a test in rock will be carried out with a pre-bored instrument deployed in pockets of slightly greater diameter than the probe, typically made with rotary coring techniques.

If the rock is weak to very weak, it may be possible to self-bore into it (Clarke et al., 1989). It is likely that this will be material from which it is difficult to recover intact core.

Coring is not essential. Destructive drilling methods can be used with some success because for the pressuremeter test it is the quality of the cavity wall that is important, not the quality of the core.

The ideal is a cored or drilled pocket no more than 5% greater in diameter than the at-rest diameter of the probe but this is an aspiration. The reality is that pressuremeter tests are often targeted at precisely those horizons where it is difficult to recover intact core and hence laboratory testing is not possible. To some extent an oversize pocket is to be expected. A successful test can be achieved in a pocket 15% to 20% greater in size than the probe but the poorer the fit, the greater the chance of membrane rupture terminating the test prematurely.

Cleanliness of the pocket is of equal importance to the diameter. If the cavity wall is lined with soft cuttings, then this can seriously degrade the test (Fig. 7.1). In this example, the probe was placed at the bottom of an apparently close-fitting pocket. The following then occurs:

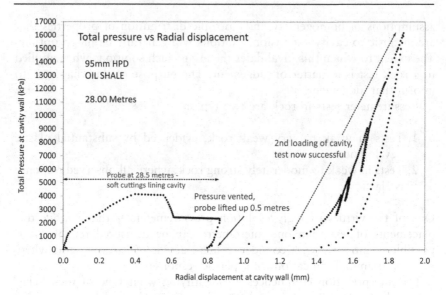

Figure 7.1 The influence of soft cuttings.

1. Pressure in the probe is raised to 4,000 kPa and is held for 2 minutes before an unload/reload cycle is attempted.
2. Despite pressure being released, the displacement continues to increase.
3. At the bottom of the failed cycle, the pressure is held for 8 minutes but the displacement still increases. This is characteristic of soft material between the probe and the cavity wall in a relatively impermeable material being extruded axially because it cannot be forced into the formation.
4. The pressure is completely released and the probe is pulled up 0.5 metres in the same pocket.
5. The test is restarted. Immediately it is apparent that the cavity diameter has increased by 2 millimetres and the test proceeds in a conventional manner. An unload/reload cycle at 4,000 KPa is now completely successful.

The test pocket needs to be long enough for a sump below the instrument to collect any residue of cuttings.

7.4 TEST PROCEDURE

Although there is an argument that the cavity should be loaded at a constant rate of strain this only applies to material that has reached the shear stress limit and thereafter is deforming plastically. Prior to this, it is more important that time-dependent deformations are allowed to develop, so tests in rock are run as a series of pressure increments, each held for a constant period (typically 1 minute). See Fig. 7.2, where the change of

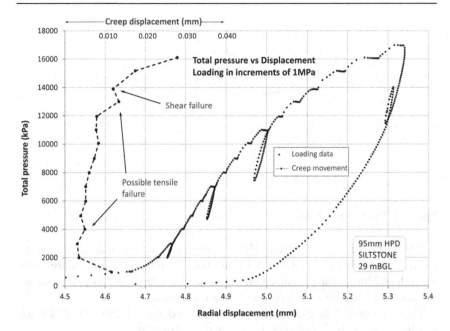

Figure 7.2 Loading the cavity with stress increments—creep response.

displacement between the start and finish of a pressure hold is plotted alongside the main loading curve on a × 10 scale. These displacements are referred to as "creep" movement. The creep curve potentially can clarify where tensile failure starts and finishes, and eventually shear failure commences. Tensile failure is recognised as a momentary increase in the creep displacement followed by a return to the previous creep rate. Once the material fails in shear, the successive steps show greater displacement as the creep rate increases.

7.5 THE ROCK MODEL

The pressuremeter test in rock can be considered from several aspects; two will be considered here.

7.5.1 Proof load test

In this approach, the pressuremeter can be treated as a large jack that forces the walls of the borehole apart. The pressuremeter being considered has an expanding section that can be envisaged as two half cylindrical plates 100 mm wide and 600 mm long, pushing against the rock, as shown in Fig. 7.3. If viewed in this way, the instrument can apply a large force of about 120 tonnes on the borehole wall.

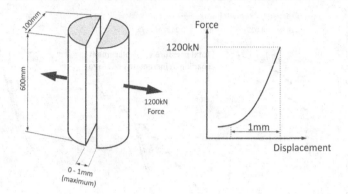

Figure 7.3 The proof load test.

When testing rock with the HPD, a typical maximum deformation or separation of these "plates" would be about 1 millimetre and is normally much less. Hence, the pressuremeter tests can be interpreted as providing an indication of the gross behaviour of the rock mass.

These gross displacements can be translated into a "modulus" by assuming that the rock behaves as an elastic continuum.

This procedure is analogous to the empirical procedure developed by Menard for determining the deformation modulus. It is only useful for comparative purposes. The actual "elastic modulus" could well be several times greater.

7.5.2 The determination of material properties from pressuremeter tests

In view of the complex nature of the rock, a simple material model of rock behaviour is proposed. This attempts to take into account the major material properties. Some limitations of this model are considered, and it is used to give some understanding of the behaviour revealed by the tests. There are six variations of the model.

7.5.3 Type (1) model: homogeneous/elastic

Consider that the rock is dry, homogeneous and behaves in an elastic manner. The stress path followed by an element of the rock adjacent to the cavity is given in Fig. 7.4 and the pressure/strain curve for this assumed material is shown for increasing and decreasing pressure alongside.

The *in situ* lateral stress is represented by point p_o. The radial stress, initially zero, increases at the same rate as the circumferential stress decreases, regardless of whether the rock is deforming under plane strain or plane stress conditions.

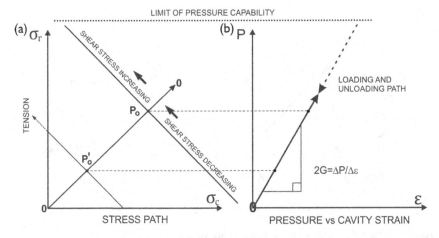

Figure 7.4 Homogeneous/Elastic.

As the pressure in the pressuremeter increases, the shear stress in the rock at the boundary wall will change, reducing at first up to the line O – O (the line containing the *in situ* lateral stress), then increasing. If the lateral stress is low, as represented by point p'$_o$, then the circumferential stress will go into tension. Assuming that the rock does not fail in tension, the loading and unloading pressure/expansion curves will be straight lines.

The shear modulus of the rock can be determined from the slope of the pressure/expansion curve using the classic procedure of Bishop et al. (1945), but little else can be derived.

7.5.4 Type (2) model: elastic behaviour before failing in shear

In this next modification, the effect of material strength is introduced. In this model it is assumed that there is a limit to the shear stress the rock can sustain. This limit is shown diagrammatically in Fig. 7.5. The rock is assumed to behave as an elastic material prior to failing in shear.

As the pressure exceeds p_f, the pressure required to initiate shear failure, the strain rate will show a substantial increase. The pressure/expansion curve that results from this model is shown alongside.

The form of the non-linear portion of the pressure/expansion curve will be a function of the shear strength of the rock. If the shear stress limit is constant and not influenced by pressure, and the rock deforms at constant volume, then the failure shear strength can be determined by the analytical procedure developed by Gibson and Anderson (1961). In practice, at least one of these assumptions will not be appropriate and a more complex solution will be required (i.e. Carter et al., 1986) (Fig. 7.6).

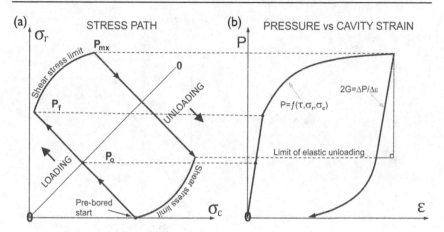

Figure 7.5 Elastic response before failing in shear only.

On unloading, the material will behave elastically until the failure strength is reached. At this point, the circumferential stress becomes the major principal stress.

In the ideal test shown in Fig. 7.5, the elastic shear modulus can be evaluated from both the slope of the loading curve (up to the onset of yield) and the unloading curve. In an actual test, it is unlikely that the true shear modulus can be determined from the loading phase. The *in situ* behaviour of the rock is often masked by the disturbance caused by the forming of the test cavity and soft cuttings being lodged between the pressuremeter and the cavity wall.

The limitation of this model is the assumption that the rock only fails in shear and not in tension. If the tensile strength of the rock is low, then the lateral stress must be high enough to ensure that the rock does not fail in

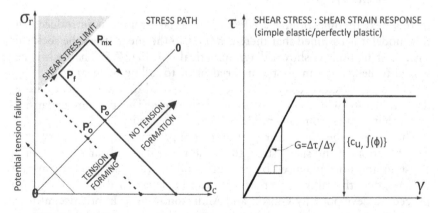

Figure 7.6 The linear elastic/perfectly plastic model.

tension and crack; otherwise, the analytical procedures developed for this model will be inappropriate.

This model represents the behaviour of some very weak rocks and all soils.

7.5.5 Type 3, 4, 5 and 6 models: elastic, shear and tension

Most rocks do not have isotropic strength properties and they are usually weak in tension and much stronger in compression or shear. If the *in situ* lateral stress is low and the shear strength high, then tensile stresses can develop during the test. The stress path followed during such a test is given in Fig. 7.7.

As the pressure increases, the stress path will follow from X to A to B. At B, the circumferential stress will start to go into tension. At some stress the rock will crack, theoretically in a radial direction (as illustrated in Fig. 7.7). With further increase in pressure, the stress path will shoot from C to D as soon as the radial crack forms. As the pressure increases, the crack will grow and the stress path will continue from D to the maximum pressure applied.

The pressure/expansion curve will show a distinct step at the initiation of the cracks. From then on, as the pressure increases and the cracks spread, the slope of the pressure/expansion curve flattens. On unloading, the stress path will follow down the same curve until the cracks close completely; then the curve will follow the initial "elastic path," as shown in the pressure/strain plot of Fig. 7.7.

If the pressure continues to increase, the mechanism of failure at the boundary will change to one of a shear failure mode as the stress path intersects the shear stress limit. Hence, in the material around the pressuremeter, there will be three distinct zones, illustrated in Fig. 7.9. (Fig. 7.8)

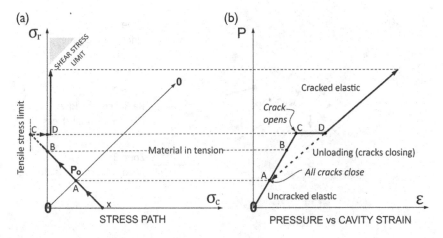

Figure 7.7 Elastic response followed by tensile failure.

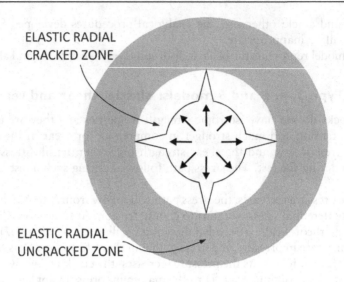

Figure 7.8 The onset of radial cracking.

- An outer zone, zone 1, in which the rock is uncracked and behaves elastically.
- Zone 2, where the circumferential stress is low. In this zone, the rock will probably not be deforming at constant volume because of the presence of open cracks.
- An inner zone, zone 3, immediately adjacent to the borehole wall. Here the rock will be failing in shear, and the voids will be closed.

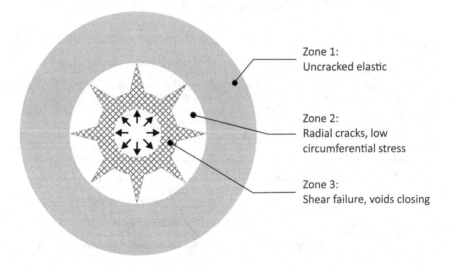

Figure 7.9 The continuation of cracking.

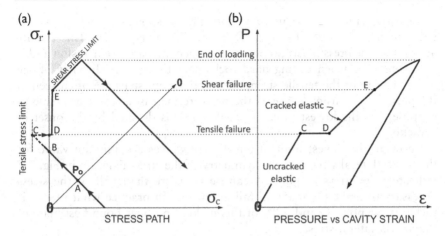

Fig 7.10 Tensile failure followed by failure in shear.

The pressure/expansion curve will have the form shown in Fig. 7.10.

If prior to the test there are existing radial cracks, then no tension can develop. The resulting stress path is shown in Fig. 7.11. If the shear strength of the rock is too high to fail under the maximum pressure of the instrument, then the pressure/expansion curve will have the form shown alongside. There is a distinct change of curvature when the crack starts to open. If the pressure is reduced, the curve will retrace itself because the material is behaving elastically at all times. However, the conventional method for determining the shear modulus will only be appropriate for the initial part of the curve, not the upper section.

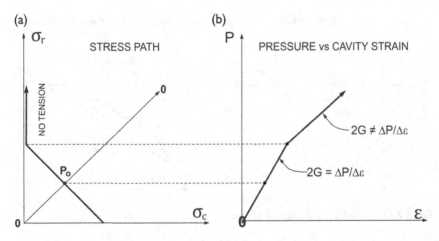

Figure 7.11 The influence of existing radial cracking.

If existing cracking does not dominate, then as soon as a crack forms, the stress path will shoot from C to D in Fig. 7.7 or Fig. 7.10. This usually appears on the pressure/strain plot as a sudden increase in expansion and is often observed when testing hard rocks near the surface. In view of the erratic nature of the tensile strength several abrupt steps are often seen in the pressure/expansion curve as the pressure increases. It is reasonable to suppose that the lowest limit of tensile stress is denoted by the onset of "cracking."

Knowing the lowest limit of tensile stress gives a maximum value for the *in situ* lateral stress. As the symmetry of the stress diagram in Fig. 7.7 indicates, the stress at point A can be no more than half of the stress necessary to make the material fail in tension. In practice what is seen is the stress at point C and 50% of this makes an upper-bound estimate of the *in situ* lateral stress.

The pressure/expansion curve will have the form shown in Fig. 7.12. It will be a straight line containing small steps but with no significant increase in strain. These steps can become closely spaced and be mistaken for shear failure, but this cannot occur until all tensile failure has ceased. On unloading, the curve will follow a similar path to the final part of the loading except that it will be smooth, without steps.

However, if on loading the cracks become filled with rock cuttings, then the unloading curve will be more stiff than the loading curve. One could expect a pressure/expansion curve of the form given in Fig. 7.13. In addition, both tension and shear failure may occur with the cracks being filled before unloading.

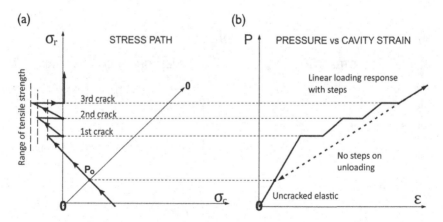

Figure 7.12 The effect of multiple fractures.

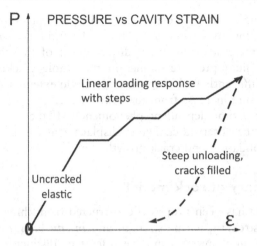

Figure 7.13 Crack filling.

7.5.6 Influence of fluid in the hole

Throughout the description, the hole is assumed to be dry during the test. Water can exist either as a consequence of the drilling process or from the hydrostatic water pressure.

If the hydrostatic water table is near the surface, then pressure will exist throughout the formation. In this situation, the previous discussions still apply with the provision that the stresses considered must be effective rather than total stresses. Fig. 7.5 is shown redrawn in terms of effective stress in Fig. 7.14.

If the rock is of low permeability (in terms of drilling time) and the natural water table is below the bottom of the hole, a very different

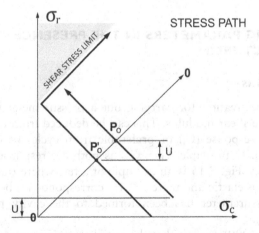

Figure 7.14 The effective stress path.

situation can occur. If in the test the hydrostatic pressure is below the pressure of the column of fluid in the borehole, and if the rock is fractured in tension by the instrument, then displacement of the borehole wall is augmented by fluid pressure acting on the whole crack. If these fluid pressures are sufficiently high then the crack could extend itself without any additional pressure increase from the instrument.

This will be a time-dependent phenomenon. If the pressures are held constant, then the observed change of displacement will be a combination of creep, consolidation and crack growth.

7.5.7 Summary of rock model

A simple model has been proposed constructed from three material parameters: shear strength, tensile strength and *in situ* lateral stress. It is apparent that the simple model can have at least six distinct forms. These are shown Fig. 7.0 and are listed below:

1. Elastic only
2. Elastic and shear failure only (which implies high lateral stress)
3. Elastic and no tension due to pre-existing fractures (low lateral stress)
4. Elastic and tension failure (low lateral stress)
5. Elastic and tension failure with crack filling (low lateral stress)
6. Elastic, tension and shear failure, with crack filling (low lateral stress).

These characteristic pressure:strain responses provide a context from which to consider pressuremeter tests in rock. The principle use of this visual model is to aid in judging when a particular analysis is appropriate, and when it is not.

7.6 DERIVING PARAMETERS IN THE PRESENCE OF FRACTURES

7.6.1 Modulus

The primary motivation for carrying out a pressuremeter test in rock is determining the shear modulus. This can be deduced from the slope of the initial loading response (G_i) or preferable from cycles of unloading and reloading (G_{ur}). In principle, $G_i \approx G_{ur}$ as both are responses to the same physical process. Fig. 7.15 is an example of a test where this is the case.

All response is elastic and there is a 1:1 correspondence between the two (once the pressuremeter has been formed to the cavity profile and all fractures close).

Ervin et al. (1980) suggest that the initial slope is indicative of the mass rock behaviour, whereas the unload/reload modulus is a measure of the

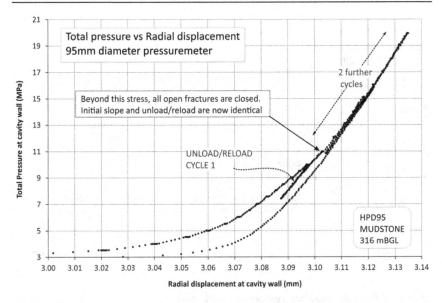

Figure 7.15 Test in mudstone—elastic only (Type I).

intact rock. It is more the case that the initial slope is affected by open jointing due to stress concentration at the cavity wall after pocket formation and the consequent unloading. It is the unload/reload modulus that is representative of the rock mass. Unlike the soil situation, it is not unusual for an unfractured rock sample loaded under laboratory conditions to give the highest modulus of all, that of the intact rock.

If the rock has not failed in shear before taking an unload/reload cycle then the ratio G_{ur}/G_i can be considered an index property describing the propensity to fracturing. If it has suffered shear failure then the ratio is more indicative of disturbance, or the degree to which the strength of the material is able to support an open cavity.

Following shear failure, if sufficient cycles are carried out, then the stress and strain dependency of modulus can be extracted from a test using relationships described in Chapter 2.

If the material does not fail, then the strain dependency of stiffness does not apply—it will be linear elastic. The stress dependency remains and it is possible to use this to make predictions of the initial stress state.

7.6.2 Shear strength

Ervin et al. (1980) give a simplified version of the Gibson and Anderson (1961) solution for cavity expansion in undrained conditions and suggest this can be used to find the shear strength of rock. It is rare to find an example of rock showing undrained behaviour. What is occasionally encountered is very weak rock that at some point in the loading deforms at a constant area ratio.

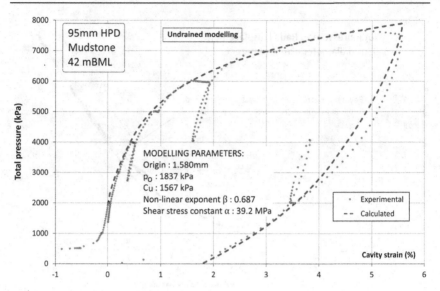

Figure 7.16 Possible undrained response in mudstone.

Fig. 7.16 shows a test in mudstone that potentially gives an undrained response. The model parameters have been derived using the methods presented in Chapter 4.

The result appears reasonable but is misleading. In particular, the cavity reference pressure, p_o, seems high when compared to the overburden stress. If a $c' - \phi'$ model is applied to the same data, the "fit" is, if anything, slightly poorer but the p_o is almost halved.

A possible explanation comes with an examination of the stiffness/strain data from the unload/reload cycles (Fig. 7.17).

Successive cycles are giving less stiff results, indicating the influence of fracture development. The steps of creep in Fig. 7.16 have been ignored by the model and so are softening the loading response. It happens that the lateral steps of strain for no pressure increase are cancelling out what would be an increasing shear stress for an increasing normal load. This response occurs when pre-existing fractures are being extended.

From a practical perspective, the values of stiffness and shear strength would be adequate for design purposes but no further parameters can be taken from the analysis. For understanding the initial stress state and the ultimate limit pressure the undrained model results are improbable. Based on the model the limit pressure would be about 11.5 MPa. Based on the loading curve, the test is almost at a limit already because fracture development is becoming the dominant element in controlling the response.

If the rock is showing dilatant behaviour, then applying an undrained solution will greatly over-estimate the strength. Ervin and Hughes consider this and judge that what is required is a $c' - \phi'$ solution, which at the time of their writing was not available. Carter et al. (1986) produced such a solution for

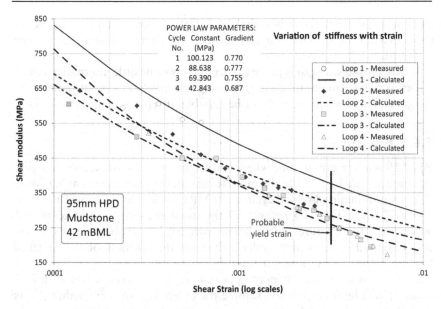

Figure 7.17 Stiffness/strain in fractured mudstone.

soils and this was modified by Haberfield & Johnston (1990) for the case of rocks. Apart from a rearrangement of terms, the two solutions are identical, so for all practical purposes the drained model used to produce Fig. 7.18 is applicable to tests where the rock has negligible tensile strength. This is a test in

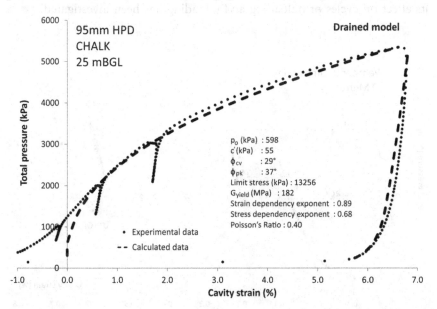

Figure 7.18 Test in chalk—shear failure only (Type 2).

chalk where the material fails in shear at a relatively low strain and techniques developed in Chapter 5 for drained tests in $c' - \phi'$ material allow the curve to be modelled.

Where tensile strength is present, there seems to be little alternative but to resort to finite element modelling or being selective about what parameters to quote. This limited set will not be capable of reproducing the field data.

7.6.3 Fracturing

Fig. 7.19 is a test in limestone where tensile failure is unmistakeable. The first occurrence is at 4.5 MPa, and a more substantial fracture occurs at 8.1 MPa.

The second unload/reload cycle at 10 MPa is a less stiff response than the first at 7 MPa. This is a reversal of the usual situation and indicates the influence of the fractures on apparent stiffness. It is probable that the first unload/reload cycle is also under-estimating the stiffness of the material because it too follows a major fracture.

As the test continues, it may appear that the material starts to fail in shear. It will be necessary to examine the creep data to see whether this is shear failure, where each step of creep is greater than the previous or merely accumulated steps of displacement associated with tensile failure (the creep movement will be constant or even reduce).

It is evident from the form of the final unloading that crack filling has taken place.

Fig. 7.20 is a test in siltstone where the fracture behaviour of the rock and its effect on cycles of unloading and reloading has been investigated.

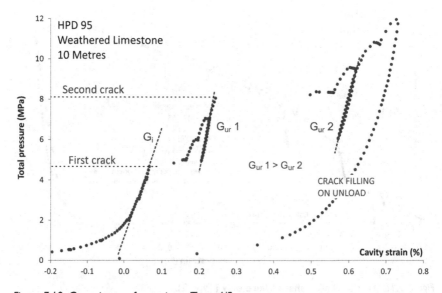

Figure 7.19 Conspicuous fracturing—Type 4/5.

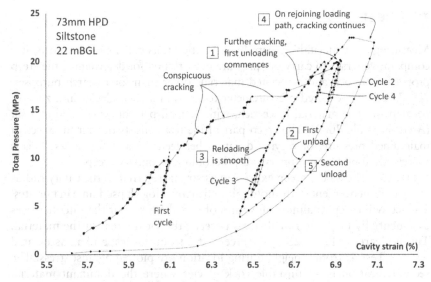

Figure 7.20 Exploring the tensile behaviour of siltstone.

1. The loading is applied in regular increments of pressure, each held for 30 seconds.
2. At 10 MPa an unload/reload cycle is taken, and the loading continues to 20 MPa.
3. Beyond 12 MPa there is extensive fracturing, successive cracks exhibiting a larger displacement.
4. At 20 MPa an unload/reload cycle is taken and gives a less stiff response than the first cycle.
5. The pressure is then released. The resulting cavity contraction goes down to 3 MPa and is smooth.
6. The cavity is reloaded to 25 MPa, with a further cycle taken when the pressure in the cavity equals that of the first cycle. The apparent reduction in stiffness is clear.
7. The reloading is smooth. There are no indications of fracturing.
8. At 20 MPa a further unload/reload cycle is taken, starting from the same pressure as that of the second cycle.
9. Between 20 MPa and 25 MPa fracturing recommences.
10. The final cavity contraction is completely smooth.

This observed behaviour bears out the predictions of the rock model. At no point in the loading does the material fail in shear. The presence of fractures makes the usual calculation of shear modulus inappropriate because of volumetric strains arising from the opening and closing of fissures. The best estimate that can be obtained is that from the very first cycle carried out at a radial stress of 10 MPa.

7.7 CREEP

Measuring creep inevitably means long-duration tests with associated costs. It is comparatively rare to carry out pressuremeter testing for determining the creep properties of rock and associated materials other than for research purposes. Fig. 7.21 is a pre-bored pressuremeter test within a rock glacier in a zone of permafrost. The material is ice rich, with a small percentage of sand and silt (Arenson et al., 2003). The creep part of the test consists of four instances of maintained pressure, of varying duration between 1.5 and 5.5 hours. This is sufficiently long to see primary creep becoming secondary creep.

Fig. 7.22 shows the creep expressed as strain (referred to the cavity radius at the commencement of the hold) plotted against log elapsed time in minutes. The ice will creep at almost all values of radial stress, but the rate increases exponentially once the radial stress exceeds the yield stress of the material. This is clearer in Fig. 7.23 where creep rates from this test and an associated test (F6 Test 4) from a neighbouring location are plotted against stress. The second location is an unstable rock glacier where the dominant material fraction is broken rock. Levels of creep in the fragmented rock are lower than in the ice until the material yields but the paths followed are similar.

A logarithmic trend gives a fit to an individual data set with correlation coefficients between 0.98 and 0.99. The maintained pressure holds for the test in ice are presented in two ways. The open circles are ultimate values with no adjustment for duration; the closed circles are the rate after a fixed time has elapsed. The improvement in the correlation coefficient shows that the

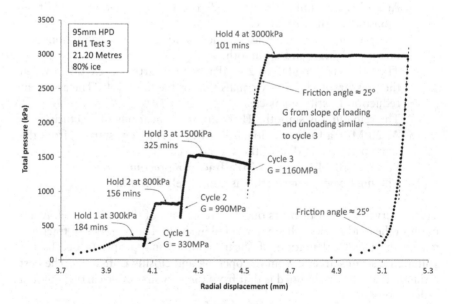

Figure 7.21 Testing for creep within a rock glacier.

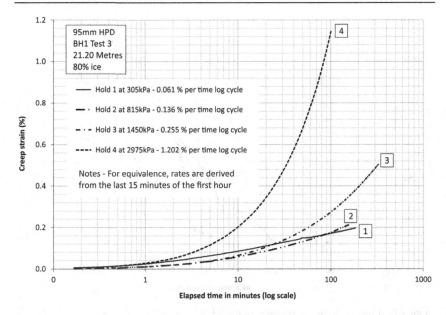

Figure 7.22 Creep strain against log time.

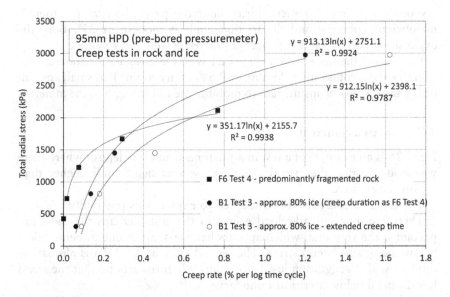

Figure 7.23 Creep rate against stress.

observed time ought to be the same for all creep stages. However, the creep duration variation does not seem to influence the next creep observation.

Values for shear modulus have been derived from the test in Fig. 7.21. The modulus is stress dependent, implying that the shearing is causing volumetric strains and hence it is appropriate to analyse the loading

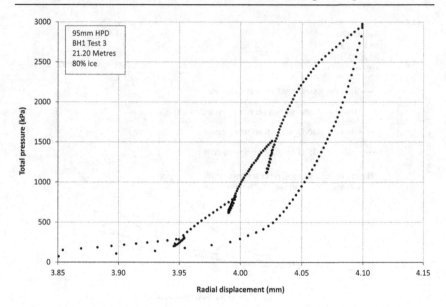

Figure 7.24 Loading curve with creep removed.

response for an angle of internal friction. In this particular example, there is no obvious cohesion and the friction angle is modest, at 25°. Both the expansion and contraction phases of the test gave similar values.

Fig. 7.24 is the same data as Fig. 7.21 with the creep data removed. The overall movement is only 0.25 mm, or 0.5% cavity strain. This small amount of expansion may be insufficient to mobilise the full strength of the material.

7.7.1 Improvisation

Fig. 7.25 is an example of a test in a siltstone/sandstone deposit where there was evidence that the lateral *in situ* stress was significantly higher than previous estimates.

The test procedure has been adapted to explore this possibility.

There are nine unload/reload events in this test. Perhaps the most important is the very first, which appears banana-shaped, due to being taken below the *in situ* lateral stress. The second cycle at 3.5 MPa is almost orthodox, with a suggestion just before the cycle turns around that the stress has dropped below a critical boundary.

Cycles 3 –7 are very regular and could be used to define a stress dependency relationship for the shear modulus (refer to Chapter 2).

Strictly, cycle 8 is not a cycle but an unloading down to a similar stress level as the start of cycle 2. The cavity is reloaded to just past the previous maximum. The cavity is then unloaded to a stress close to the turn-around point of cycle 2. The pressure is held, and the creep movements monitored.

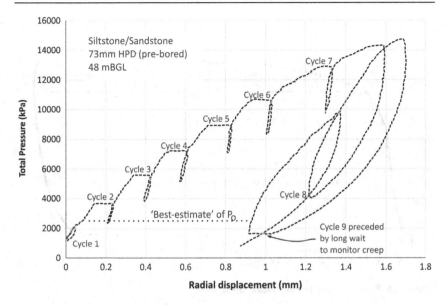

Figure 7.25 Investigating Po using unload/reload cycles and creep.

The stress is well above the overburden but there is a large inward movement, from which it can be surmised that this stress is below the insitu lateral stress.

Cycle 9 then follows and, on its return down to zero, the path falls within the envelope of the original unloading. Assembling all this evidence points to a value for the *in situ* lateral stress that seems to coincide with an inflexion in the initial loading. The lateral stress is nearly three times the vertical.

7.8 PORE COLLAPSE

Fig. 7.26 is a self-bored test in chalk from a platform in shallow sea water. At approximately 5 MPa and shortly after the material shows indications of shear failure, the structure of the chalk breaks down, a phenomenon termed "pore collapse." It happens in reservoir rocks and is usually associated with increased effective stress due to hydrocarbon extraction and an abrupt compaction of the rock mass. It is comparatively rare to see it in a pressuremeter test because the sensible operator will have terminated the test as soon as accelerated movement was evident. Here, the test pressure was controlled by an electronic system, keeping the cavity strain rate reasonably uniform. A final contraction curve mirrors the initial loading path.

The mechanism is the opposite of the hydrocarbon extraction situation. The material is highly porous with water flowing freely through the

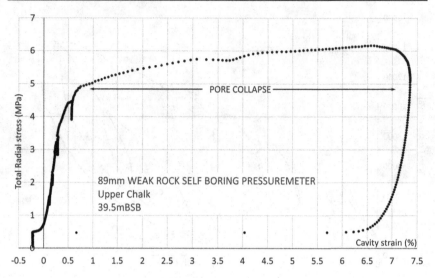

Figure 7.26 Pore collapse in very weak chalk.

formation. The effective stress is increasing at the same rate as the total stress. Once shear failure has led to a small amount of compaction, the porosity of the rock reduces and excess pore pressures are generated. This lowers the effective stress and the compaction is able to continue with negligible increase of total stress. The process is similar to liquefaction.

7.9   THE DIFFICULTY OF FINDING COMPARATIVE DATA

Fig. 7.27 illustrates the general problem faced by the engineer who requires parameters for a difficult material.

The figure consists of two images. On the left is a photograph of a 0.5 metre long length of core. No piece of core is longer than about 75 mm. On the right is the test that was made at the centre of the displayed core. The test itself is a complete loading and contraction, going to a reasonably high strain and all cycles are showing a similar response. 90 MPa is the lowest shear modulus, from the slope of the first cycle. The initial slope is less than a third of this.

At just below 2 MPa there are the first indications that the material is showing shear failure. Hence from this test, credible and representative values for stiffness and strength can be obtained despite the lack of intact core. Without such core, laboratory testing cannot be carried out, hence there may be no comparable alternatives to the pressuremeter test.

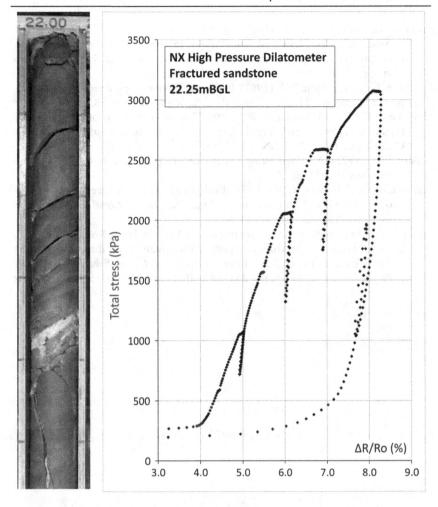

Figure 7.27 Core and test for a fractured sandstone.

REFERENCES

Arenson, L.U., Springman, S.M. & Hawkins, P.G. (2003) Pressuremeter tests within an active rock glacier in the Swiss Alps. *Proc. 8th International Conference on Permafrost*, Zurich. CRC Press, ISBN 9789058096906.

Bishop, R.F., Hill, R., & Mott, N.F. (1945) The theory of indentation and hardness tests. Proceedings of the Physical Society **57** (3), pp. 147–159.

Carter, J.P., Booker, J.R., & Yeung, S.K. (1986) Cavity expansion in cohesive frictional soil. *Géotechnique* **36** (3), pp. 349–358.

Clarke, B.G., Newman, R. & Allan, P.G. (1989) Experience with a new high pressure self-boring pressuremeter in weak rock. *Ground Engng* **22** (6), pp. 45–52.

Ervin, M.C., Burman, B.C. & Hughes, J.M.O. (1980) The use of a high capacity pressuremeter for design of foundations in medium strength rock. *Proc. International Conference on Structural Foundations on Rock*, Sydney. 7–9th May, pp. 9–16.

Gibson, R.E., & Anderson, W.F. (1961). In situ measurement of soil properties with the pressuremeter. Civil Engineering and Public Works Review, 56, pp. 615–618.

Haberfield, C.M. & Johnston, L.W. (1990) The interpretation of pressuremeter tests in weak rock – theoretical analysis. *Proc. 3rd Int. Symp. Pressuremeter*, Oxford, pp. 169–178. ISBN 0727715569.

Ladanyi, B. (1972) An engineering theory of creep of frozen soils. *Canadian Geotechnical* 9 (1), pp. 63–80.

Ladanyi, B. & Johnston, G.H. (1973) Evaluation of in situ creep properties of frozen soils with the pressuremeter. *Proc. 2nd International Conference on Permafrost*, Yakutsk, pp. 310–318.

Springman, S.M., Arenson, L.U., Yamamoto, Y., Maurer, H.R., Kos, A., Buchli, T. & Derungs, G. (2012) Multi disciplinary investigations on three rock glaciers in the Swiss Alps: Legacies and future perspectives. *Geografiska Annaler Series A Physical Geography* 94, pp. 215–243. 10.1111/j.1468-0459.2012.00464.x

Chapter 8

Miscellaneous Experiments with Pressuremeters

(experimental use of the cavity expansion test)

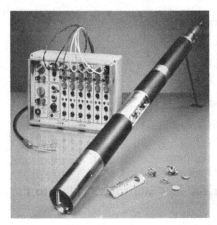

Figure 8.1 Six-axis load cell pressuremeter and controller.

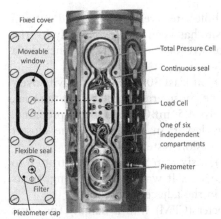

Fixed cover

Moveable window

Total Pressure Cell

Continuous seal

Load Cell

One of six independent compartments

Flexible seal

Piezometer

Filter

Piezometer cap

Figure 8.2 LCPM transducer compartment.

DOI: 10.1201/9781003200680-8

199

8.1 NOTATION

α	Shear stress constant (Bolton & Whittle, 1999)
β	Exponent of non-linearity (Bolton & Whittle, 1999)
ε	Cavity strain
E	Elastic (Young's) modulus
G	Shear modulus
k_o	Co-efficient of earth pressure at rest
P	Total pressure measured at the cavity wall (radial stress)
P_o	Initial cavity reference stress (total)
p'_o	Initial cavity reference stress (effective)
P_{lim}	Total limit pressure
σ_r	Radial stress
σ_θ	Circumferential stress (also termed hoop stress)
r	Radius
r_f	Radius of cavity at first yield
r_c	Current radius of cavity
U	Pore water pressure
γ	Shear strain
υ	Poisson's ratio (0.5 for the undrained case)

8.1.1 Introduction

This chapter considers some of the fringe uses and trials applied to the pressuremeter and its test.

8.2 MEASURING THE EXTENT OF THE PLASTIC ZONE

In May 2016, an examination of borehole interference was carried out at a test location in Cambridgeshire. The site has been used extensively to trial and demonstrate pressuremeter technology. Ratnam et al. (2005) includes a description of its geotechnical characteristics.

The material from 4.5 metres down to at least 30 metres is a heavily over-consolidated deposit of Gault clay. Some considerable time before the trial, two boreholes cased to approximately 4.5 mBGL had been prepared. Unusually, the borehole spacing was less than 0.5 metres, the closest achievable using a standard cable percussion system.

In the first of these boreholes, BH14, a six-arm SBP self-bored from 6 to 7 metres, giving a test centre at 6.5 metres. It was left monitoring, so its piezometers would respond to events in the adjacent borehole.

In position BH18, a cone pressuremeter (CPM) with a dummy cone was pushed from 6 to 7 metres, so that the centre of its inflation point was in a similar plane to that of the SBP. A quick undrained test was carried out with the CPM. This was followed within a few minutes by a similar test with the SBP.

The SBP self-bored another metre and carried out a further test. The situation is sketched in Fig. 8.3 and the field curves in Fig. 8.4. The expansion is plotted as radial displacement for maximum definition but expressed as cavity strain, the CPM test has expanded four times further than the SBP tests.

The CPM test allows the shear strain at first yield to be derived, γ_y, in this case 0.645%, equivalent to a rigidity index of 155. Because the pushed installation and subsequent test takes the material to limit pressure, the radial extent of the elastic-plastic boundary R is directly proportional to the cavity radius r_c:-

$$R = r_c(\gamma_y)^{-0.5} \tag{8.1}$$

Using the result for γ_y, the radial extent has been derived for the following:

- Installation of the CPM
- Testing with the CPM
- Testing with the SBP

These boundaries are shown in Fig. 8.3.

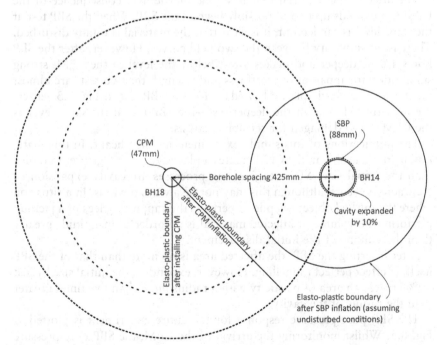

Figure 8.3 Interfering boreholes layout.

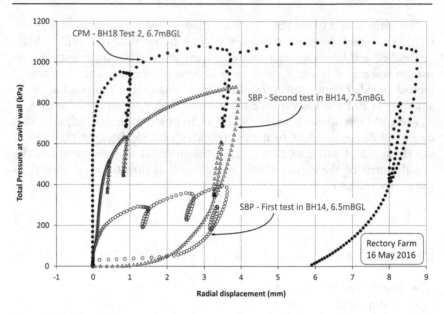

Figure 8.4 Field curves.

The area of the material that has gone plastic as a consequence of the CPM test extends up to and possibly beyond the SBP. When the SBP test at the same level is carried out, it is clear that the material is totally disturbed. There is no similarity between the two field curves. However, after the SBP bores a little deeper and makes a test below the level of the CPM, strong similarities are apparent (for stiffness and strength the two tests are almost identical). The elastic-plastic boundary for the SBP test in Fig. 8.3 reflects the undisturbed case for the deeper test—the SBP test at the same level as the CPM is too damaged for sensible analysis.

The implications of this simple experiment are significant. In this material, merely pushing in the CPM creates a plastic zone 12.5 greater in radius than the initial size. Many engineering processes are cavity expansion or contraction events, although this may not be always obvious. In a situation where the pushed object is a pile or perhaps the supporting legs of a jack-up platform, the volume of affected material is an order of magnitude greater than the volume of the initial displacement.

After inserting the SBP, the affected area is no more than that of the SBP itself, in effect perfect tunnelling. However, expanding the initial size by just 10% creates an area of plasticity with a radius more than five times greater than the radius of the cavity.

The SBP pore pressure response for the entire experiment is plotted in Fig. 8.5. Whilst monitoring the arrival of the CPM, the SBP pore pressure transducers show a small rise. Calculations suggest this is the sub-yield elastic response of the material. Once the failed zone passes through the

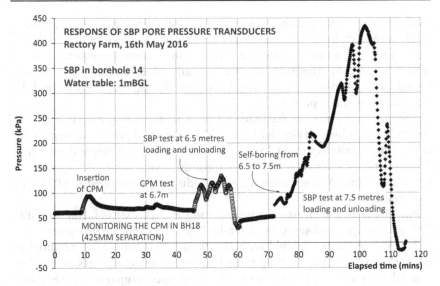

Figure 8.5 Pore water pressure response.

material occupied by the SBP, there are only minor observable fluctuations attributable to the expansion phase of the CPM test.

Having been failed, the response to further shearing suggests that the material dilates and consequently the path followed by the first SBP test is largely drained. It is possible to see in Fig. 8.4 that the membrane collapses at a head of water, suggesting a free-draining material.

For the deeper SBP test, the response is as expected from an undrained loading and unloading.

8.3 THE SUB-YIELD EXPERIMENT

An earlier version of the previous experiment had been carried out in 1997. In this case, the motivation was demonstrating that the non-linear elastic/ perfectly plastic solution of Bolton & Whittle (1999) gave accurate predictions of the elastic response at a remote distance.

Two boreholes were prepared, with 0.42 metres between centres. Down the first, a load cell pressuremeter (LCPM) self-bored from 9.5 to 10.5 m, giving a test centre at 10 metres. After a few hours, its readings settled to a constant value. An expansion self-boring pressuremeter (SBP) then bored from 9.5 to 10.5 metres in the adjacent borehole.

Because of the time taken by the LCPM test, it is only occasionally utilised, generally for research purposes. The self-boring method is used to position a multi-transducer load cell at a specified depth. The external stress acts on stiff but movable plates bonded to load cells. As a consequence of this stress, the plates begin to move inwards. A feedback system sees the load cell output change, and compensates by applying internal gas pressure to the probe, making the plates move back until a null position is restored. The internal pressure is now the same as the external pressure. Over a period of time, the stresses arrive at a constant value, in principle the *in situ* lateral stress. Additional transducers measure the internal pressure and the external pore water pressure.

The current version of the load cell pressuremeter is shown in Figs. 8.1 and 8.2. An older device built in 1975 was used for these tests with two diametrically opposed load cells seeing the same internal pressure. The null position is therefore a compromise, with both cells showing a small deflection. A single pore water transducer is positioned at 90° to the load cells.

For this trial, the stress cells were arranged with Cell C facing the SBP. Cell A is on the opposite side of the LCPM, so 81 mm further away.

The SBP test was analysed for strength and the shear strain required to make the material yield. Using these results, the stress changes at a radius centred on the LCPM have been calculated and compared to the actual readings of the LCPM whilst the SBP expansion test was ongoing.

Fig. 8.6 shows the results of the self-bored expansion pressuremeter test. Fig. 8.7 gives the outputs from the load cells and the predicted output based on the SBP results. The expansion achieved with the SBP was enough to make the material around the load cell shear elastically but not quite enough to reach the fully plastic state.

The prediction and the measured data are remarkably close. A number of conclusions can be drawn. Compared to a linear elastic/perfectly plastic solution, Bolton and Whittle predict a higher yield stress and some pore

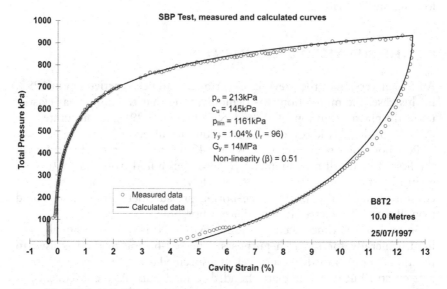

$p_o = 213\text{kPa}$
$c_u = 145\text{kPa}$
$p_{lim} = 1161\text{kPa}$
$\gamma_y = 1.04\%\ (I_r = 96)$
$G_y = 14\text{MPa}$
Non-linearity (β) = 0.51

○ Measured data
— Calculated data

B8T2
10.0 Metres
25/07/1997

Figure 8.6 SBP (sub-yield response experiment).

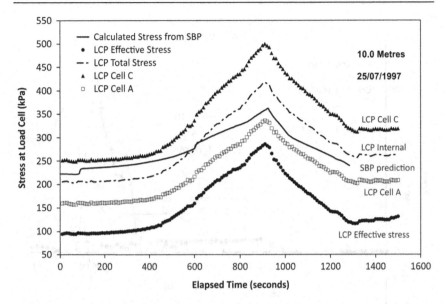

Figure 8.7 LCP (sub-yield response experiment).

water generation in the sub-yield response. Both of these turned out to be the case and of the correct magnitude.

The data also show why self-boring is the preferred installation method whenever possible. The LCPM, prior to the SBP test commencing, was reading a similar initial stress value to that obtained from the SBP test analysis.

8.4 MEASURING THE ELASTIC MODULUS WITH THE LOAD CELL PRESSUREMETER

The current production version of the LCPM has six hermetically isolated compartments, each containing a load cell, piezometer and internal pressure cell (Fig. 8.1 and Fig. 8.2). A complex pneumatic controller regulates the internal pressure inside each compartment independently of its companions.

Referring to Fig. 8.2, the central part of the cover is a plate screwed down to the load cell. During the process of finding the null state, the controller does not settle on the "right" value immediately. The internal pressure can be greater or less than the external stress as the controller slightly overshoots and then falls back. When this occurs, the stiff plate that covers the load cell is carrying out an unload/reload event at very small strain. An example is given in Fig. 8.8. The circled points are the data of interest.

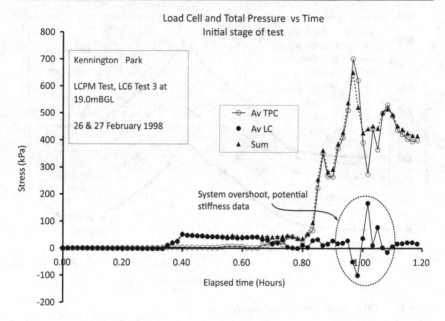

Figure 8.8 LCPM at Kennington Park.

This test was part of project to investigate the state of stress around an old underground tunnel beneath Kennington Park in London (Gourvenec et al., 2005). Location LC6 was 19.5 metres from the centre line of the tunnel. At about the same distance and 4 metres away was a self-bored expansion pressuremeter (SBPM) location, EP6. At a similar spacing, a rotary cored hole provided samples for laboratory testing.

The LCPM plates do not make a continuous surface; consequently, their movement compresses rather than shears the ground. The relevant modulus is the elastic modulus in the horizontal direction, E_h. The stress-strain characteristics of the load cells are known. A speculative calculation can be carried out assuming the process is analogous to a circular plate load test:

$$E_h = \left(\frac{P}{m}\right)\left(\frac{\pi}{4}\right)D(1 - v_H^2)I_f \qquad [8.2]$$

where
  P is the external load
  m is the resulting movement
  D is the width of the plate
  v_H is Poisson's ratio in the horizontal direction (assumed 0.1)
  I_f is a combined influence or shape factor (assumed 1).

Calculation gives an E_h of 915 MPa and a G_h of 315 MPa, assuming an undrained response. If the LCPM result is applied to the data from the nearby SBPM test, the elastic threshold strain result is 2.4×10^{-5}.

Some small strain triaxial testing was carried out and the nearest result (16.4 metres) to the test in Fig. 8.8 is a shear modulus of 220 MPa. Unfortunately there is no independently determined value for the elastic modulus at the level of the test in Fig. 8.8. However, if the threshold strain is similar, then an SBPM test at 15 metres predicts a shear modulus of 186 MPa. It is sufficient agreement to show that the procedure has merit.

This is not the best example of the process, but it is the only case to date where there are direct comparative data. Figs. 8.9 and 8.10 give the result from an LCPM at the Rectory Farm test site (referenced above). In this case, the LCPM was in the ground for 3 days. After 2 days an unambiguous steady-state reading of the *in situ* lateral stress had been achieved. The internal pressure was then vented over the course of 9 hours and the external stress allowed to compress the load cell webs. The primary purpose was seeing to what extent the established *in situ* stress would be recovered after permitting some stress relief. The process led to a stress relieving of the ground by about 60 kPa. The secondary purpose was to use this stress reduction to determine the small strain modulus.

The average value from these data is equivalent to a shear modulus of 203 MPa (assuming an undrained event—there was no decline in pore water pressure during this stage of the test).

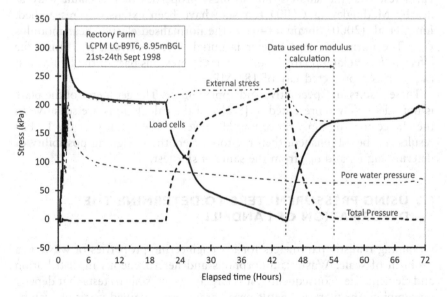

Figure 8.9 Three-day LCPM test.

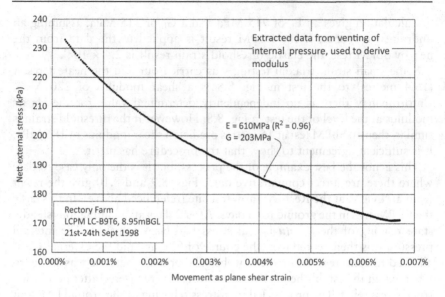

Figure 8.10 Calculating the elastic modulus.

If this result is used as a starting point for the onset of stiffness decay, then SBP tests at similar levels suggest that the elastic threshold shear strain is $\approx 1.1 \times 10^{-5}$.

Although there are no independent comparative data from the Rectory Farm test site, the small strain stiffness properties of the Gault Clay at nearby Madingley (a CUED test site) have been extensively researched. Lings et al. (2000) obtain 1140 for the normalised elastic shear modulus G_{hh}. The normalising parameter is initial effective stress, p'_o. Using the effective *in situ* lateral stress value from the previous part of the LCPM test as p'_o gives a predicted G_{hh} of 181 MPa.

These results are speculative and approximate. The geometry of the plate is not adequately represented in [8.2] and the calculations are sensitive to the choice for Poisson's ratio, ν_{hh}. However, if the validity of the LCPM results *can* be established, then the tool offers the intriguing possibility of determining E_h and σ_{ho} from the same *in situ* test.

8.5 USING PRESSUREMETERS TO DETERMINE THE DEFORMATION OF LANDFILL

Obtaining realistic parameters for the engineering properties of waste is a problem of scale. Waste is amorphous and heterogeneous in distribution and element size. Consequently, it is largely inaccessible to tests that depend on sampling the material. Settlement parameters obtained vertically can be derived from surface loading (Abbiss, 2001), but the lateral response can only be obtained by *in situ* methods. Knowing the strength and stiffness of the waste is necessary for evaluating the long-term stability and integrity of landfill lining systems, and for re-developing exhausted waste sites.

Dixon et al. (2006) report the results of a series of experiments using self-bored, pushed and pre-bored pressuremeters in municipal solid waste (MSW). The age of the waste varied from less than 1 year to more than 15 years old. In view of the hostile nature of some of the material tested, the success rate of the testing is surprisingly high. For practical reasons, the preferred technique is pre-bored testing with a high-pressure dilatometer (HPD). The high-pressure capability is not required, but the large expansion range is essential. The waste needs to undergo large shear strains in order fully to mobilise its strength.

The boreholes were prepared using a continuous flight auger system. Once near the intended test level, pockets for the HPD were formed with a barrel auger. Inspection of the barrel auger contents allowed the type of waste to be matched to the resulting test. In most cases, the HPD needed to be pushed into the pockets. Examples of the obtained field curves are given in Figs. 8.11, 8.12 and 8.13.

All test levels are below the leachate, but ongoing decomposition means that some of the 1-year waste cannot be assumed to be fully saturated.

This is an engineered waste and as it ages it acquires some cohesion. The form of the test is indistinguishable from a test in free draining soil and it can be analysed using the $c' - \phi'$ analyses for drained expansion presented in Chapter 5. Stiffness is highly non-linear, with a strong dependency on stress level (Dixon et al., 2006). The imposed shear strain in the examples is greater than 35% and all tests are very close to a limit state at the end of loading.

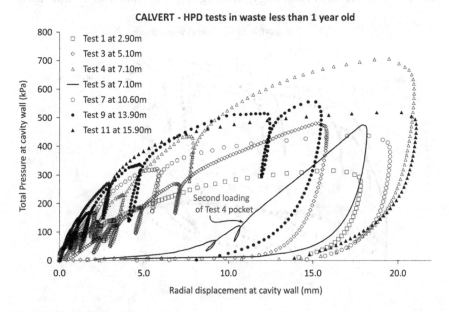

Figure 8.11 One-year-old waste.

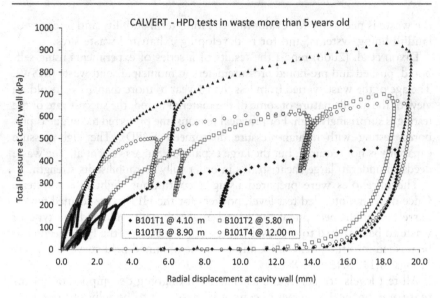

Figure 8.12 Five-year-old waste.

Several of the tests were repeated after a delay of several hours and without removal of the probe from the ground. There is an example in Fig. 8.11.

Fig. 8.14 is an example of waste that has not been formally engineered. The location is an area of shallow waste deposition on the fringes of the Indian Institute of Technology, Kharagpur. The material is unusually

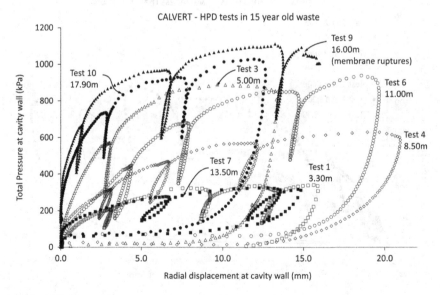

Figure 8.13 Fifteen-year-old waste.

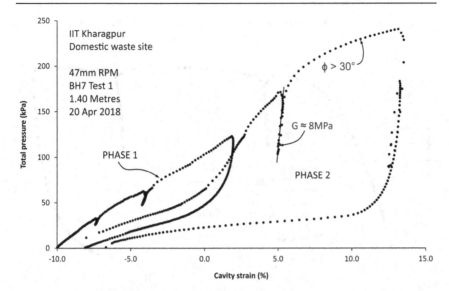

Figure 8.14 Unengineered waste.

well-sorted and consists almost entirely of shredded paper and plastics. Pockets were made with a 50 mm hand auger and the pressuremeter was a small reaming pressuremeter (RPM) being deployed as a pre-bored device.

The test is in two parts. The waste is unconfined and in places under consolidated. The material is not a contiguous mass and the first 10% of the expansion finds the point where true shearing commences. Once established, the cavity is unloaded and reloaded to the displacement limit of the equipment. This is sufficiently large to see that the deformation is approaching a limiting state. The strength and stiffness is surprisingly high considering the shallow depth.

8.6 PULVERISED FUEL ASH (PFA)

Fig. 8.15 is an example of a pushed pressuremeter test in pulverised fuel ash (PFA), a significant waste product of coal-fired power stations. The waste is deposited in settling lagoons and allowed to consolidate. PFA is frequently used as filling material in concrete production.

The example was likely to have been deposited about 15 years earlier. The test gives an unambiguous undrained response with significant stiffness and strength.

Typically its engineering properties have been decided by testing with a cone penetrometer (CPT), but the correlations required cannot be taken from existing values for natural ground (Stewart et al., 2006). The pressuremeter used for the example was a 15 cm^2 cone pressuremeter (CPM)

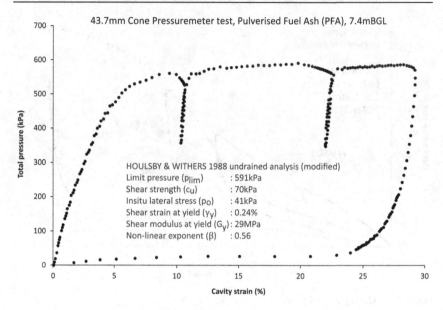

Figure 8.15 Pushed test in pulverised fuel waste.

with a live cone on the front end and was pushed from a cone truck. For this material, it is the optimum arrangement of techniques and gives a means of deriving site specific correlations for further CPT tests.

8.7 STRESSES AROUND A DEEP TUNNEL

HADES is an underground research laboratory located 223 metres below surface at the Belgium Nuclear Research Centre (SCK CEN) near Mol. It is a set of tunnels (nominally 4 metres diameter) constructed approximately at the centre of a 110-metre thick deposit of Boom Clay. Its primary purpose is investigating the potential for the geological disposal of Category A nuclear waste.

Experimental pressuremeter testing in the clay immediately surrounding the tunnel was carried out soon after the initial test drift was completed (Clarke & Allan, 1990). After further construction, a programme of self-bored and pre-bored pressuremeter testing was carried out between 1999 and 2009. These were along and across the tunnel alignment, the latter boreholes being horizontal, vertical and angled.

The material is a poorly indurated (friable) clay and at these depths can be considered a soft rock. It has novel sealing characteristics that mean any hole made in the material closes in a relatively short time and eventually heals—that is, all evidence of a prior fracture is completely erased. This behaviour is accelerated if water is added so consequently the *post*-1990 testing was carried out using air as the flushing fluid to minimise changes to the insitu properties of the clay.

Fig. 8.16 is an example from a series of horizontal tests carried out in 1999. The test is 30 metres in advance of the tunnel face, the objective being to obtain parameters for the undisturbed clay.

The deployed instrument is a six-axis self-boring pressuremeter fitted with an inclinometer. Opposing pairs of displacement followers have been averaged to give three readings at 120° spacing with axis 1 almost vertical. The figure indicates that axis 1 sees less movement than the other two, these being similar to each other. The inference is that the *in situ* vertical stress is higher than the lateral and hence k_o is less than 1.

An "average" analysis of the combined output using the curve modelling approach outlined in Chapter 4 gives a cavity reference pressure of 4,078 kPa. As this is lower than the best estimate of the total vertical stress of 4,375 kPa it is likely to be the *in situ* lateral stress. The ambient pore water pressure is 2,177 kPa giving a k_o of 0.86.

The calculation of k_o cannot be done using data from the initial part of the loading curve. Although the vertical stress is the highest, the initial movements in the vertical direction are the greatest. This apparent contradiction is caused by the need to drill fractionally over-size in order to bore the material without needing to mobilise unrealistic jacking forces. The overcut is about 1.5% greater than the diameter over the measuring section and allows some small movement of the instrument axis with respect to the borehole axis.

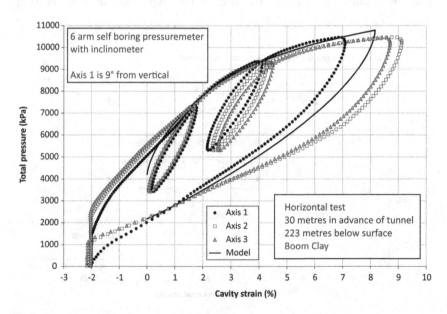

Figure 8.16 Self-bored test (horizontally orientated).

However, when unload/reload cycles are taken, the probe is locked in place by the high levels of stress being applied. Using the output of the multiple displacement sensors, it is possible to see the variation in modulus across three diameters at 120°. This variation can be used as input for a Mohr's circle calculation. It cannot compute the absolute value of initial stress, but can provide the ratio of major to minor stress and orientation. Additional checks have to be carried out to ensure that what is being seen is a ground and not an instrument effect as the spatial data variation is small. Chapter 3 provides a description of this method.

For the test shown in Fig. 8.16, the resulting ratio of minor over major stress is 0.91 and this can be taken to be an alternative estimate of k_o. This is derived with no requirement for data for pore water pressure or vertical stress. Using the inclinometer data, the major stress is determined to be acting at 174°. As this is close to vertical, the difference probably represents the degree of uncertainty in the experimental data.

Fig. 8.17 shows data from additional horizontally orientated self-bored pressuremeter tests taken closer to the tunnel wall. At 6 metres into the material, the *in situ* lateral stress has risen to the levels seen in the 1999 data, which is the benchmark representing the far field response.

All these tests showed substantial excess pore water pressure being generated during the loading phase.

The carrying out and interpretation of the test is made more complex by the tendency of the material to seal itself. In Fig. 8.16, it is evident that despite the over-cut that has been applied to the boring of the cavity, the material has closed onto the body of the probe. The expansion only starts after

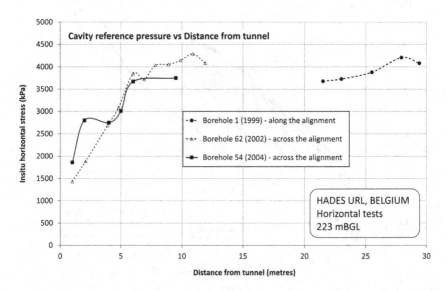

Figure 8.17 Initial lateral stress and distance from tunnel.

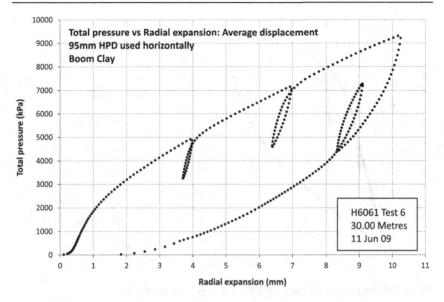

Figure 8.18 A pre-bored pressuremeter test in a horizontal borehole.

considerable stress has been applied to the cavity wall. This is a characteristic that is accelerated by horizontal orientation in material with anisotropy in the initial stress conditions. In 2009, a series of horizontally orientated *pre-bored* pressuremeter tests were attempted at the same location. Fig. 8.18 is an example for a test at a similar distance from the tunnel wall as the self-bored test shown in Fig. 8.16. The consequences of the rapid closure of the material are considerably more severe for the pre-bored test in a pocket that was initially bored at 102 mm. Some analysis is possible, but the majority of the loading data are discarded in the evaluation of parameters. This borehole closed onto the probe in a matter of minutes.

If the borehole is orientated vertically, then the problem is minimised, at least for the length of time it takes to drill a pocket and perform the test. Because the horizontal axis is now positioned in an isotropic stress field, borehole closure takes several hours. The borehole wall is loaded by the pressuremeter before this movement can occur and a more convincing field curve is obtained where almost all the data are following the same stress:strain path. Self-bored and pre-bored tests are virtually indistinguishable (Fig. 8.19).

As part of the evaluation of the legitimacy of the Mohr's circle calculation using unload/reload data, the vertical tests were also examined using the same procedure. For these tests, the variation between axes was less than 3%. This indicates isotropic conditions and defines the degree of experimental uncertainty that can be expected using this method.

Some of the descriptions and data in this account are taken from an unpublished paper by Bidwell et al. (2001).

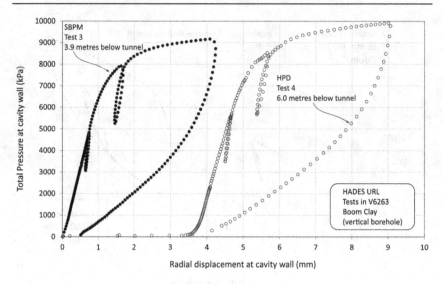

Figure 8.19 Successive SBP and HPD tests in a vertical borehole.

REFERENCES

Abbiss, C.P. (2001) Deformation of landfill from measurements of shear wave velocity and damping. *Géotechnique* **51** (6), pp. 483–492.

Bidwell, T.J., Bernier, F. & Buyens, M. (2001) Development of the self-boring pressuremeter for testing in deep clay. *Unpublished* but available on request from Cambridge Insitu Ltd. Email cam@cambridge-insitu.com.

Bolton M.D. & Whittle R.W. (1999) A non-linear elastic/perfectly plastic analysis for plane strain undrained expansion tests. *Géotechnique* **49** (1), pp. 133–141.

Clarke, B.G. & Allan, P.G. (1990) Self-boring pressuremeter tests from a gallery at 220 m below ground. Proc 3rd Int. Symp "Pressuremeters", Thomas Telford Ltd., pp 73–84. ISBN 0 7277 1556 9.

Dixon, N., Whittle, R.W., Jones, D.R.V. & Ng'ambi, S. (2006) Pressuremeter tests in municipal solid waste: measurement of shear stiffness. *Géotechnique* **56** (3), pp. 211–222.

Gourvenec, S.M., Mair, R.J., Bolton, M.D. & Soga, K. (2005) Ground conditions around an old tunnel in London Clay. Proc. ICE, Geotechnical Engineering 158, Iss. GE1, pp. 25–33.

Lings, M.L., Pennington, D.S. & Nash, D.F. (2000) Anisotropic stiffness parameters and their measurement in a stiff natural clay. *Géotechnique* **50** (2), pp. 109–125.

Ratnam, S., Soga, K. & Whittle, R.W., (2005) A field permeability measurement technique using a conventional self-boring pressuremeter. *Géotechnique* **55**, pp. 527–537.

Stewart, D.I., Cousens, T.W. & Charles-Cruz, C.A. (2006) The interpretation of CPT data from hydraulically placed pfa. *Engineering Geology* **85** (1–2), pp. 184–196. ISSN 0013-7952.

Epilogue

This book combines the experience and knowledge acquired by the two authors and on that basis has taken about 100 years to write.

We have picked a narrow path through the plethora of pressuremeter technology and methods used to produce values for design purposes. Our prediction is that ultimately only one approach will endure—it will be scrupulous measurement coupled with fundamental analysis.

The standard penetration test (SPT) and the cone penetrometer test (CPT) are the most commonly used *in situ* testing tools available to site investigation engineers. We are supporters (and occasional users) of both—in their place. The SPT gives one number at any particular depth. The CPT gives one number at any particular depth in sands and potentially up to three numbers in clay. None of these values are measurements of displacement and consequently are indeterminate. Considerable effort has been put into finding relationships between these numbers and engineering parameters—it is an exercise worthy of the mathematicians of Laputa.[1]

One device capable of measuring movement has been available for the same length of time as the modern CPT. The original Ménard Pressuremeter was a work of genius—a simple tool that made coarse measurements of the pressure and movement of the expanding membrane. Fundamental solutions for such data existed almost from the start, but the resolution of the equipment has never been what is required to fulfil this potential. Consequently, a conservative practice has developed around the Ménard probe that is dependent on empiricism.

Empiricism makes life simple, arguably safe and is costly. It has to be this way. The purpose of an empirical number is to cover an expanse of uncertainty with a single value. At best, designs based on parameters decided empirically will be over-cautious. For a certain kind of engineering problem where it is essential to *know* how much movement is going to occur they are worse than useless. All numerical modelling falls into this demanding category of problems where it is important to input representative and accurate data. The model will decide the appropriate safety level once the values are shown to give realistic predictions.

217

The approach to analysis outlined in this book insists on the essential coherence between measured data and parameter set. If the analysis process is selective and partial, then this coherence breaks down. In principle and in fact, given the parameters, the location of every single measured data point can be predicted. Other than matching field performance it is the nearest it is possible to get in this industry to a proof. This is the potency of cavity expansion solutions and is the aspect that empiricism abandons.

There have been a few false dawns for the equipment described in this text due to over-simplified analysis but occasionally the converse. Ground engineering practice, faced with reconciling two very different answers for the same material, tends to play safe. It will increase the number of hoops for the newer and therefore "riskier" technique and will be inclined towards consensus. There is only one test of the data that counts and that is the post-construction response of finished structures. The high-resolution pressuremeter has a proven ability to get the prediction right and is often the only technique that does so. We would argue that given the measured response of the structure and the pressuremeter data, any disagreement between the two is a function of the model used to predict the displacements, not the pressuremeter parameter used as input.

To be clear — the data provided by these tools are always accurate but sometimes our understanding is deficient. We have at several instances in this book gone back to data that are more than 30 years, sometimes 50 years old and re-interpreted them in the light of better analytical tools. From the beginning, when advanced pressuremeters started to come out of the Cambridge University Engineering Department, the measurements of movement have been good enough to see micrometres of change. That is the key to comprehending the behaviour of the ground because that allows us to obtain realistic shear modulus through small stress cycling.

Unloading and reloading the ground is an extraordinarily powerful process yet simple to understand and interpret. It is only necessary to be able to *see* the rebound event. For that, you need to measure something very small.

We are at an interesting point. The high-resolution pressuremeter is mature technology. It can always be improved but is now close to being as reliable as it can be, whilst acknowledging that from time to time the ground will destroy the outer cover and fill the device with water!

These are complex instruments. They demand a lot from the operator and data analyst, and that will continue to be the case. Accept this complexity in order to benefit from it. Sophistication by itself, of course, is not the entire answer—the CPT is immeasurably more sophisticated than the SPT but as far as the ground is concerned they are both pointed sticks.

It would be a mistake to think that all development and improvement in the probe has come from the manufacturer or inventor. Many talented people with no commercial incentive have thought enough of the potential of the tools to devote time towards enhancing their performance. For example,

Fahey and Jewell[2] (1990) consider the issue of instrument compliance. Their improvements to the displacement measuring system have been a major advance in the testing of stiff soils.

Our understanding of the cavity expansion test is probably now adequate—for two particular stress paths and one highly simplified stress/strain response. There are many materials and conditions that do not fall into these categories, unsaturated soils being one. Better analyses will come. Better empirical factors for the SPT will not.

However, these pressuremeters could characterise a site at a few key points and then the CPT or possibly the SPT might be able to fill in the gaps. That is the intelligent use of the strengths of both types of testing. Depressingly, what tends to happen is the reverse, with the pressuremeter being asked to explain why its values don't agree with the CPT. We hope this book answers that question.

ACKNOWLEDGEMENTS

If you steal from one author it's plagiarism. If you steal from many its research.

Wilson Mizner

There are many individuals who have been persuaded by the potential of the advanced pressuremeter and have applied their intellect into developing the understanding of the test. Most of them will be found by examining the references.

Special mention is due to John "Clive" Dalton. He started Cambridge Insitu specifically to make a commercial product from the prototype self-boring pressuremeters built by John Hughes. His early appreciation of John's work shows remarkable foresight from someone whose background is not soil mechanics. It took many years before his faith was justified by the returns.

Clive's co-director for most of this time was Philip Hawkins, a gifted mathematician and engineer, especially in electronics. Philip designed and developed the ancillary equipment for controlling the test and capturing the data. He has those desirable attributes, an aptitude for field work and an intense curiosity about what the data mean. The method of describing the sub-yield non-linear stiffness of sands was first worked out by Philip.

Of the many talented people who worked for or alongside Cambridge Insitu, two made a major contribution to the development of the pressuremeter. Tim Bidwell, who never let a poor argument pass unchallenged, and the late Patrick Finlay. Pat was a genius of the lathe but was so much more. He had been John's technician at Cambridge

University. Of the current people at CI, grateful thanks to Yasmin Byrne, whose encouragement and willingness to take on obscure calculations has been invaluable.

There are very few ideas (and certainly none in this book) whose development is solely down to the efforts of one person. Unreserved apologies to all at Cambridge Insitu and elsewhere for their contributions which have been incorporated into this text but inevitably will be unacknowledged.

CI have helped and been helped by several PhD students and graduates from Cambridge University. Dr Sang Ratnam and Dr Liu Lian ("Leo") significantly advanced the understanding and were excellent company on site. The brilliant Ken Sutherland was another, who never had the chance to reach his potential.

Over the years, we have received generous support and encouragement from Professor Kenichi Soga and Professor Malcom Bolton. Thanks to Dr Richard Newman, not only for his enthusiasm for high quality pressuremeter testing over many years but also for his considered, if somewhat depressing, assessment of where we are at the moment with this technology[3].

Maxime Bourgouin, formerly of Fugro UK, put considerable time into checking much of the text and equations: It is chastening to have your grammar corrected by someone for whom English is their second language.

Thanks to Professor Debasis Roy of IIT Kharagpur and Professor Renato Cunha of the University of Brazil, both former PhD students at UBC. Dwayne Smith devised the means of testing the complex gravels described in Appendix B. Thanks to Keith Brown for his continuing support for the U.S. pressuremeter operation conducted under the auspices of In Situ Engineering, Washington.

This book would not have been possible without Elizabeth Carmack keeping one of the authors alive and restoring him to something approaching full functionality. It was Elizabeth who suggested using quinoa seed for the CPT experiments in Appendix D.

DATA

Cambridge Insitu Ltd, in the course of developing the pressuremeters designed by John Hughes, carried out a substantial amount of field trials and participated in a number of research projects. These have provided the bulk of the examples reported in this book.

Other sources include the London Crossrail project, particularly the data from the 1992 investigations, and the work carried out by SCK CEN investigating options for storing nuclear waste.

There is a substantial body of published data for these pressuremeters in a variety of journals. Géotechnique has been a significant source. Where we have used such information it is acknowledged in the text.

We appreciate the generosity of Professor Sarah Springman for allowing access to field data from work carried out by ETH, Zurich (Figure E.1).

Figure E.1 Clive and Simon—former and current managing director of Cambridge Insitu Ltd.

NOTES

1 Jonathan Swift's *Gulliver's Travels*.
2 Fahey, M. and Jewell, R. (1990). Effect of pressuremeter compliance on measurement of shear modulus. Proc. ISP.3 Oxford, ISBN 0 7277 1556 9, pp. 115–124.
3 Private communication. His assessment of how engineers use, or rather don't use, sophisticated data has guided our approach to this book.

Analytical Basis of the Pressuremeter Test

The stages of cavity expansion and contraction

Fig. A.1 to A.5.

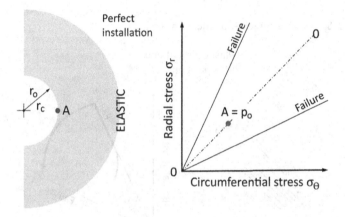

Figure A.1 Stress development—initial state.

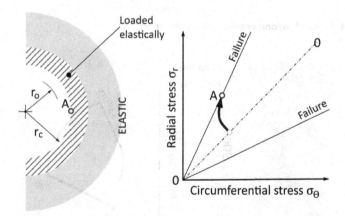

Figure A.2 Stress development—elastic expansion.

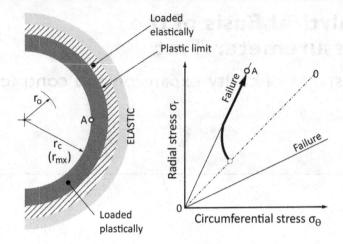

Figure A.3 Stress development—plastic expansion.

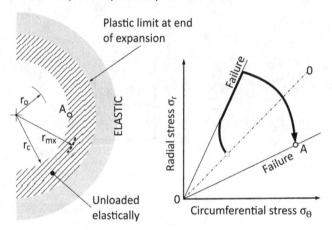

Figure A.4 Stress development—elastic contraction.

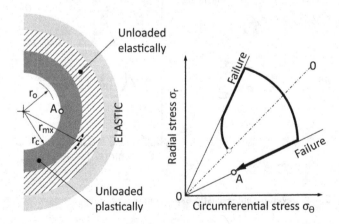

Figure A.5 Stress development—plastic contraction.

A.I NOTATION

a_o	Initial radius of the cavity
a	Radius of the cavity wall
ξ	Finite displacement – ξ_a is displacement at the cavity wall
c_u	Undrained shear strength
G	Shear modulus
E	Young's modulus
v	Poisson's ratio
τ	Shear stress
p, p_{limit}	Pressure measured at the cavity wall, limit pressure at infinite strain
p_o	Cavity reference pressure (potentially the insitu lateral stress, σ_{ho})
r, r_a	Radius to a point in the material and the radius of the cavity wall
ε_θ, ε_r, ε_z	Circumferential strain, radial strain, vertical strain
ε_c, ε_{ip}	Circumferential strain at the cavity wall and at the initiation of plasticity
γ, γ_c, γ_y	Shear strain, measured at the borehole wall, to reach first failure
σ_r, σ_θ, σ_z	Radial stress, circumferential stress, vertical stress
σ_1, σ_3	Principal stresses
σ'_{av}	Mean effective stress
u, u_o	Pore water pressure, excess and ambient
S, (S_{ul})	Slope of the loading (unloading) response when drained field data are plotted on log – log scales.
ϕ'	is angle of shearing resistance. ϕ'_{cv} is the friction angle when the material is being shearing at constant volume
ψ	is dilation angle
c'	is drained cohesion
M	is $\dfrac{(1 + sin\psi)}{(1 - sin\psi)}$
N	is $\dfrac{(1 + sin\phi')}{(1 - sin\phi')}$
R	Radius of the elastic/plastic transition

A.2 PLANE STRAIN CAVITY LOADING

The purpose of this appendix is to describe how the cavity expansion and contraction solutions are derived from the measurements of radial displacement and radial stress at the cavity wall. Only the most essential elements of the analyses are presented. The relevant chapters on drained and undrained strength further develop the solutions and their application.

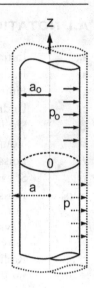

Figure A.6 Axial symmetry.

The pressuremeter test is the loading of a long cylindrical cavity where the axis of the cylinder is vertical. If the height of the cylinder is considerably greater than its diameter the cavity can be treated as infinitely long. Consequently, there are no movements in the vertical direction; it is only necessary to consider stresses and strains in the horizontal plane. Fig. A.6 shows the cavity expanded from an initial radius of a_0 to a greater radius a. The stress required to do this increases from a reference stress p_0 to p.

Because it is a plane strain and stress configuration the unknowns are the radial and circumferential stresses, σ_r and σ_θ, and the radial and circumferential strains, ε_r and ε_θ. The analytical problem is to use the pressuremeter measurements to derive all four unknowns for all points in the surrounding mass of soil.

A.2.1 Elastic strains

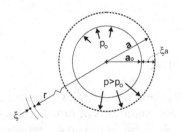

Initial conditions:

$\sigma_r = \sigma_\theta = p_0$

Initial cavity diameter = $2a_0$

Expanded conditions:

Pressure at cavity wall = p

Displacement at cavity boundary a is ξ_a

Displacement in soil mass at radius r is ξ

Figure A.7 A horizontal section through the cavity.

Consider the case of a pressuremeter perfectly installed in an infinite soil mass (Fig. A.7). This is a horizontal section through the expanding cavity. The starting position of the cavity wall (synonymous with the outer surface of the pressuremeter membrane) is shown by the inner solid circle. The starting pressure being applied equals the cavity reference pressure, p_o. The initial radius of the cavity is a_o. As the applied pressure increases ($p > p_o$), the cavity expands a distance, ξ_a, to the dotted circle, radius a. A remote point in the soil mass, initially at radius r will move by a smaller amount, ξ.

If a cylindrical cavity in an isotropic medium is loaded elastically, then deformations take place at constant volume (Timoshenko & Goodier, 1934), which for a plane strain event reduces to constant area.

The area of material displaced at the inside boundary where the displacement is ξ_a must equal that area which moves a distance ξ at a radius r:

$$\pi(a_o + \xi_a)^2 - \pi a_o^2 = \pi(r + \xi)^2 - \pi r^2$$
$$\therefore 2a_o\xi_a + \xi_a^2 = 2r\xi + \xi^2$$

For small strains, ξ_a^2 and ξ^2 can be ignored, leading to:

$$\xi = \xi_a \frac{a_o}{r} \qquad\qquad\qquad [a.1]$$

Hence, displacements are inversely proportional to the radius and the decrease of radial displacement ξ with radius is defined throughout the soil mass by [a.1].

A.2.2 Shear strains

The shear strains resulting from this radial displacement can be calculated by considering what occurs to a small square element of the soil, ABCD, located at radius r with sides of unit length. This element is deformed to position A*B*C*D* as the diagonal DB moves a distance ξ to D*B* (Fig. A.8).

Because ABCD is of unit length, DB is $\sqrt{2}$. As every point in the element moves radially, the increase in the length of DB in moving a distance ξ is given by:

$$D^*B^* - DB = DB\frac{(r + \xi) - r}{r}$$
$$[a.2]$$
$$\therefore D^*B^* - DB = \sqrt{2}\frac{\xi}{r}$$

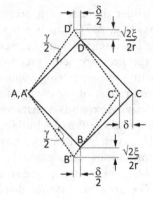

Figure A.8 An element of soil being sheared.   Figure A.9 Shear strains.

Further, as there is no change in area when the element ABCD deforms to A*B*C*D*, the change in the length of AC by δ is given by:

$$AB \times AD = \frac{1}{2}(B^*D^* \times A^*C^*) = \frac{1}{2}\left(\left(BD + \sqrt{2}\frac{\xi}{r}\right)(AC - \delta)\right)$$

$$AB = AD = 1, \quad \text{and} \quad BD = AC = \sqrt{2} \quad 1 = \frac{1}{2}\left(\left(\sqrt{2} + \sqrt{2}\frac{\xi}{r}\right)(\sqrt{2} - \delta)\right)$$

$$\therefore \sqrt{2}\delta = \frac{2\xi}{r} - \sqrt{2}\delta\left(\frac{\xi}{r}\right)$$

If the strains are very small, $\frac{\xi}{r}\delta$ can be ignored. Hence:

$$\delta = \sqrt{2}\frac{\xi}{r} \qquad\qquad\qquad [a.3]$$

The element in moving from its initial position at ABCD can be considered to have moved bodily outwards without inducing any stresses and then sheared. The shear strains can be seen by translating the deformed element back. Fig. A.9 shows the same information as Fig. A.8, but with the deformed element arranged so that A and A* coincide.

Shearing distorts the shape of the soil element. Shear strain, γ, is the difference between the angles B*A*D* and BAD in radians, and is given by:

$$\gamma = 2\frac{\xi}{r} \qquad\qquad\qquad [a.4]$$

Hence, shear strain is directly proportional to displacement. As the displacements throughout the soil mass are defined by [a.1], the shear strains are also defined and are the same both normal and tangential to the direction of load.

A.2.3 Stress equilibrium

Fig. A.10 shows the principal stresses acting on a small element being sheared as a consequence of a cavity expansion. At all times, the material is in equilibrium so the sum of the forces acting on the element equal zero. The face of the element nearest the probe is at radius r and the shearing angle is θ. Because θ is infinitesimal, it is denoted $\delta\theta$.

Using polar coordinates, the moments referred to the cavity centre can be stated. Noting that the face furthest from the cavity is at radius $r + \delta r$ and adopting the soil mechanics convention that compressive forces are positive:

$$r\sigma_r\delta\theta + 2\left[\delta r\sigma_\theta \sin\frac{\delta\theta}{2}\right] = [r + \delta r]\left[\sigma_r + \delta r\frac{d\sigma_r}{dr}\right]\delta\theta \qquad [a.5]$$

Small-angle approximation means that $\sin\delta\theta = \delta\theta$ and this allows $\delta\theta$ to be eliminated:

$$r\sigma_r + \delta r\sigma_\theta = [r + \delta r]\left[\sigma_r + \delta r\frac{d\sigma_r}{dr}\right] \qquad [a.6]$$

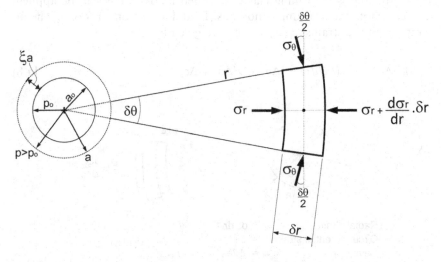

Figure A.10 The equilibrium of stress on an element of soil or rock.

Expanding the quadratic gives a squared term that disappears as the element dimensions approach zero. This leads to the equilibrium differential equation for stress:

$$- r\frac{d\sigma_r}{dr} = \sigma_r - \sigma_\theta \qquad\qquad [a.7]$$

The importance of [a.7] is that the right-hand side can be expressed in terms of the failure stresses of the ground, allowing the differential to be solved.

A.2.4 Strain equilibrium

Fig. A.11 is Fig. A.10 redrawn to show the strains on an element of soil undergoing shear. The cavity expansion from a_o to a has resulted in a radial displacement, ξ_a.

The relevant strains are:

- Radial strain ε_r is $-d\xi/dr$
- Hoop or circumferential strain, ε_θ, is change of circumference over the original circumference, hence, ξ/r
- Strain at the cavity wall, ε_c, is ξ_a/a_o. ξ_a is measured by the pressuremeter so ε_c is an easily determined case of circumferential strain.

A.2.4.1 Strain compatibility

If the material is deforming elastically, then Hooke's law can be applied. The constants are a Young's modulus, E, and a Poisson's ratio, v. The alterations in the strains, ε_r, ε_θ and ε_z, are given by:

$$E. \Delta\varepsilon_r = -E(d\xi/dr) = [\Delta\sigma_r - v(\Delta\sigma_\theta + \Delta\sigma_z)] \qquad\qquad [a.8]$$

Radial strain	$\varepsilon_r = d\xi/dr$
Circumferential strain	$\varepsilon_\theta = \xi/r$
Cavity strain	$\varepsilon_c = \xi_a/a_o$

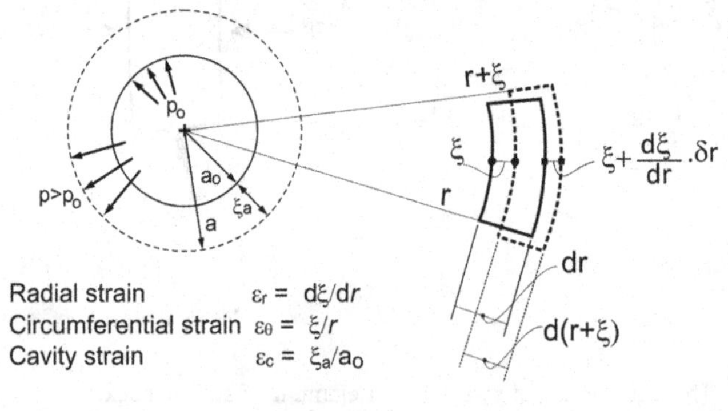

Figure A.11 Strains around an expanding cavity.

$$E. \Delta\varepsilon_\theta = -E(\xi/r) = [\Delta\sigma_\theta - \nu(\Delta\sigma_r + \Delta\sigma_z)] \qquad [a.9]$$

$$E. \Delta\varepsilon_z = [\Delta\sigma_z - \nu(\Delta\sigma_r + \Delta\sigma_\theta)] \qquad [a.10]$$

In these equations, ξ is the displacement at radius r. Because of the assumption of plane strain cavity deformation, the axial strain ε_z is 0. Hence:

$$\Delta\sigma_z = \nu(\Delta\sigma_r + \Delta\sigma_\theta) \qquad [a.11]$$

Combining these definitions eventually leads to:

$$\Delta\sigma_r = \left(\frac{-E(1-\upsilon)}{(1-2\upsilon)(1+\upsilon)} \right) \left(\frac{d\xi}{dr} + \left(\frac{\upsilon}{1-\upsilon} \right) \frac{\xi}{r} \right) \qquad [a.12]$$

$$\Delta\sigma_\theta = \left(\frac{-E(1-\upsilon)}{(1-2\upsilon)(1+\upsilon)} \right) \left(\frac{\xi}{r} + \left(\frac{\upsilon}{1-\upsilon} \right) \frac{d\xi}{dr} \right) \qquad [a.13]$$

Equ. [a.12] and [a.13] can be used to solve the stress equilibrium equation [a.7]. After some manipulation, the result is the strain equilibrium relationship:

$$r^2 \frac{d^2\xi}{dr^2} + r\frac{d\xi}{dr} = \xi \qquad [a.14]$$

A.2.5 Simple shear

Applying the above definitions, the displacement, ξ, at a radius r is given by:

$$\xi = \sigma_r(1+\upsilon)r_a^2/(r * E) \qquad [a.15]$$

At the cavity wall, σ_r is the measured pressure p. The measured displacement is ξ_a and the radius is r_a. Hence:

$$E = p(1+\nu)/(\xi_a/r_a) \qquad [a.16]$$

Shear modulus, G, and the Young's modulus, E, are related by $E = 2G(1+\nu)$, consequently:

$$G = \frac{p}{2} / \frac{\xi_a}{r_a} \qquad [a.17]$$

The elastic deformations and stress changes determined by the pressure-meter are a direct function of the shear modulus. Despite the loading being applied in a compressive direction, the material experiences pure shear.

The radial stress σ_r at any radius r as a function of the measured pressure and measured cavity radius is given by:

$$\sigma_r = p\frac{r_a^2}{r^2} \qquad\qquad\qquad [a.18]$$

Provided that the material response is linear elastic, the circumferential stress σ_θ is given by:

$$\sigma_\theta = -p\frac{r_a^2}{r^2} \qquad\qquad\qquad [a.19]$$

For the linear elastic case, $\Delta\sigma_r = -\Delta\sigma_\theta$ and the shear stress τ at any point in the plane of deformation is:

$$\tau = p\frac{r_a^2}{r^2} \qquad\qquad\qquad [a.20]$$

Whilst the material response is linear elastic, changes in the measured pressure are equal to changes in the mobilized shear stress. This will persist either until the full strength of the material is mobilized (linear elastic) or until the strain exceeds the elastic threshold (non-linear yield).

Shearing in the pressuremeter test occurs on vertical planes (Fig. A.12). Beyond the failed zone, the mobilized shear stress declines with radius. This differs from laboratory testing where for a sample being sheared in the triaxial apparatus the shear stress is constant throughout (Fig. A.13).

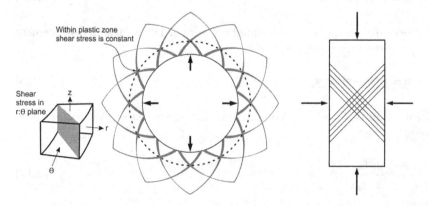

Figure A.12 Planes of shear for a cavity expansion.   Figure A.13 Triaxial shearing.

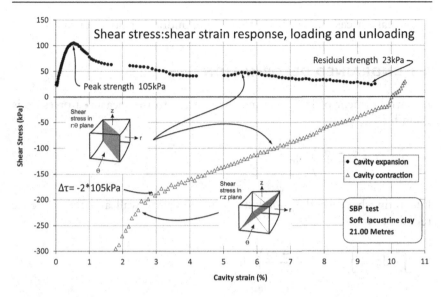

Figure A.14 Full shear stress:strain response with axes rotation.

For almost all realistic soils shear takes place on the r:θ plane. Wood & Wroth (1977) consider the possibility of shear occurring on other planes depending on the prevalent k_0 conditions. Due to non-linearity of the soil response, there is a disproportionate increase in radial stress compared to circumferential stress, increasing the likelihood of preferential failure on the r:θ plane. The possible exception is cavity *contraction* in material where k_0 is in the normal to lightly over-consolidated range. There is then a real possibility of rotation in the principal stress axes and failure transferring to the z:θ plane (Fig. A.14).

A.2.6 Initial observations

This description has derived strains from displacements and has produced the stress and strain equilibrium equations. The stress equilibrium must be satisfied for all stages of a cavity expansion and contraction and is therefore fundamental.

The strain equilibrium result is only applicable whilst displacements are directly proportional to the applied stress. It has been customary to assume this is the case prior to the full strength of the material being mobilized, in other words yielding. In practice, linear proportionality is confined to a vanishingly small stress alteration and the equation is of limited use. The concept of point of maximum shear stress remains useful but deriving sub-yield strains requires the incorporation of non-linear stiffness into the calculation of the yield coordinate.

The next stage of this description considers practical implementations of the stress equilibrium equation that in addition take account of non-linearity in the stress-strain response.

The easiest case to resolve is that of the undrained cavity expansion.

A.3 UNDRAINED DEFORMATION

A rearrangement of [a.7] relating shear stress to the measured pressure is given below:

$$d\sigma_r = -\left(\frac{\sigma_r - \sigma_\theta}{r}\right). dr$$

Integrating between the radial stress at the cavity wall and the reference radial stress:

$$\int_{p_0}^{p} d\sigma_r = \int_{r=a}^{r=\infty} -\left(\frac{\sigma_r - \sigma_\theta}{r}\right). dr$$

Hence:

$$p - p_0 = \int_{r=a}^{r=\infty} -\left(\frac{\sigma_r - \sigma_\theta}{r}\right). dr \qquad [a.21]$$

The pressure change at the cavity wall is the integrated shear stress response. If the stress-strain curve is known, then the pressure-expansion curve can be constructed. In practice, the pressure-expansion curve is measured and the stress-strain curve has to be found. At the wall, using the measured pressure and displacements, [a.21] is differentiated to find the shear stress $(\sigma_r - \sigma_\theta)/2$.

This has been done by[1] Palmer (1972) for the undrained case with respect to strain at the cavity wall, ξ_a/a_o or ε_c. This gives:

$$\sigma_r - \sigma_\theta = \frac{2\varepsilon_c dp}{d\varepsilon_c} \quad \text{or} \quad \tau = \varepsilon_c \frac{dp}{d\varepsilon_c} \qquad [a.22]$$

This is a small strain approximation. The exact solution is:

$$\tau = \frac{\varepsilon_c}{2}(1 + \varepsilon_c)(2 + \varepsilon_c)\frac{dp}{d\varepsilon_c} \qquad [a.23]$$

This may be written as:

$$\tau = \frac{dp}{d[\log_e(A/A)]} \qquad [a.24]$$

The Palmer solution applies to every part of the pressure-strain response; there is no distinction between elastic or plastic behaviour. In the form of [a.24], it is the current slope of the plot of total pressure on a linear scale against shear strain on a log scale.

An example of the Palmer solution being applied to expansion and contraction data for a test in soft clay is given in Fig. A.14. This is a normally consolidated clay. Towards the end of the final unloading, it appears that the principal stress axes rotate.

A.3.1 Shear stress from contraction data

Because the solution applies everywhere, values for the current shear stress during cavity contraction can also be derived. Changes of cavity strain and radial stress are referred to the point where the loading ends and the contraction commences but in all other respects the calculation is the same.

A.3.2 Circumferential stress

Equ. [a.22] can be rearranged to find the circumferential stress, σ_θ. If the contraction shear stress is also derived, then a plot of radial stress against circumferential stress can be produced for the entire test, as in Fig. A.16.

Inevitably the trend is "noisy." Although the source data seem reasonably quiet (Fig. A.15), in fact pressure has been injected via solenoid valves so the path followed has a slight sawtooth form. Deriving shear stress from this and subsequently deducting it from the radial stress multiplies the

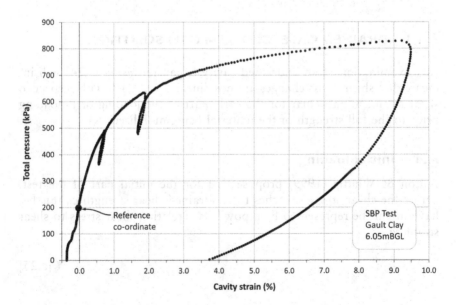

Figure A.15 Pressure—strain plot in Gault Clay.

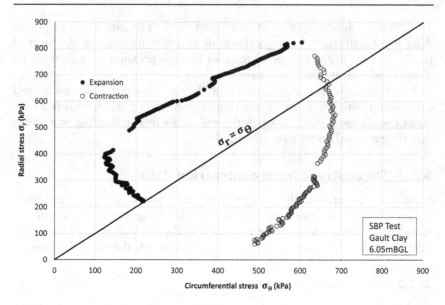

Figure A.16 Radial against circumferential stress.

noise. Nevertheless, the trend is clear and significantly non-linear, especially for the cavity contraction phase.

In principle, the effective stress path could be followed in a similar way. In practice, very few tests combine good pore pressure data and near zero insertion disturbance for this to be achievable.

A.4 UNDRAINED CASE, CLOSED-FORM SOLUTION

Any examination of an undrained cavity expansion using [a.24] will indicate that shear stress changes are non-linear (Fig. A.17) with respect to shear strain, in particular for the early part of the loading or unloading prior to the full strength of the material being mobilised.

A.4.1 Initial loading

Bolton & Whittle (1999) propose that for the initial part of the test, before the shear stress τ reaches the undrained shear strength c_u this behaviour can be represented by a power law relating shear stress to shear strain, γ:

$$\tau = \alpha\gamma^\beta \qquad\qquad [a.25]$$

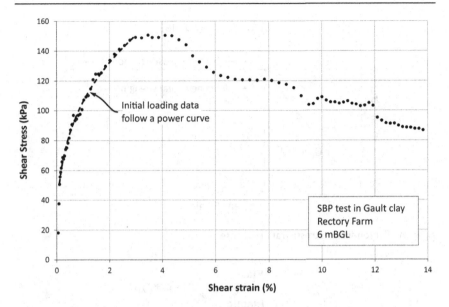

Figure A.17 Fit of power law to stress:strain response.

The power law terms are identified by drawing a power curve through unloading or reloading data obtained from rebound cycles. β is the exponent of linearity and typically lies between 0.5 and 1, where 1 is linear elastic. α is the shear stress constant and would be shear modulus for linear elasticity. Using this result, it can be shown that a solution to [a.24] whilst $\tau < c_u$ is given by:

$$p - P_o = \frac{\alpha}{\beta}\gamma^\beta \qquad\qquad [a.26]$$

Hence, failure occurs when the pressure increases above the reference stress at the cavity wall is equal to c_u/β.

Bolton and Whittle use this result to re-evaluate the well-known closed-form solution given by Gibson & Anderson (1961) for an undrained cavity expansion. Fig. A.18 is the stress-strain response assumed by Bolton and Whittle, which is non-linear until the full strength is mobilised and thereafter is perfectly plastic.

A.4.2 Plastic loading

As the expansion continues, there is a plastic zone confined by the limiting elastic radial stress of $p_o + c_u/\beta$. The stress distribution around the cavity is shown in Fig. A.19.

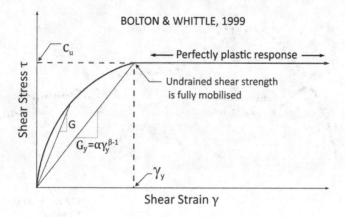

Figure A.18 Non-linear stress:strain response.

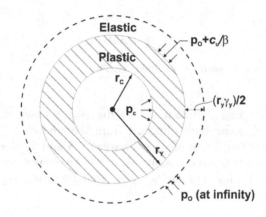

Figure A.19 Equilibrium stresses and radii.

The stress equilibrium equation [a.7] still applies and can be written as:

$$r\frac{d\sigma_r}{dr} + 2c_u = 0 \qquad \text{[a.27]}$$

This gives:

$$d\sigma_r = -2c_u\frac{dr}{r} \qquad \text{[a.28]}$$

Integrating between the radii of the cavity wall and of the elastic-plastic transition:

$$\int_p^{p_o+\frac{c_u}{\beta}} d\sigma_r = -2c_u \int_a^R \frac{dr}{r} \qquad \text{[a.29]}$$

Hence:

$$p = p_0 + c_u \left[\frac{1}{\beta} + \ln\left(\frac{R^2}{a^2}\right) \right]$$

[a.30]

In soil sheared at constant area:

$$\left(\frac{R}{a}\right)^2 = \frac{\gamma_c}{\gamma_y}$$

This leads to:

$$p = p_0 + c_u \left[\frac{1}{\beta} + \ln(y_c) - \ln(y_y) \right]$$

[a.31]

The result resembles the simple elastic/perfectly plastic solution proposed by Gibson and Anderson. For the particular case of a linear elastic response when $\beta = 1$, the two solutions are identical. Indefinite expansion of the borehole is predicted by:

$$p_{\text{limit}} = p_0 + c_u \left[\frac{1}{\beta} - \ln(y_y) \right]$$

[a.32]

Substituting in [a.31]:

$$p = p_{\text{limit}} + c_u \ln(y_c)$$

[a. 33]

This indicates that the undrained shear strength and limit pressure can be obtained from the gradient and intercept of a plot of total pressure at the cavity wall versus the natural log of the current cavity shear strain.

y_c is the shear strain at the cavity wall. y_y is the shear strain at which the material first reaches the undrained shear strength and is given by:

$$y_y = \left(\frac{c_u}{\alpha}\right)^{1/\beta}$$

[a.34]

The reciprocal of this is the rigidity index of the material.

A.4.3 Tensile stresses

Using [a.25] and the stress equilibrium equation, it follows that prior to yield:

$$\sigma_r - \sigma_\theta = 2\tau = 2\alpha\gamma^\beta$$

[a.35]

From [a.26]:

$$\sigma_r = p_o + \frac{a}{\beta}\gamma^\beta \qquad\qquad [a.36]$$

Hence:

$$\sigma_\theta = p_o - \alpha\left(2 - \frac{1}{\beta}\right)\gamma^\beta \qquad\qquad [a.37]$$

If β takes a value of about 0.5, then $\sigma_\theta \approx p_o$ and will remain so until the material becomes fully plastic. If the circumferential stress cannot reduce, then tensile stresses cannot be generated. The possibility of cracking emerges as a characteristic of a linear elastic response more appropriate for rock (Fig. A.21).

Fig. A.20 sketches the general case for the asymmetrical development of radial and circumferential stresses implied by non-linearity.

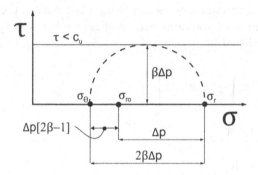

Figure A.20 Non-linear stress development (from Bolton & Whittle, 1999).

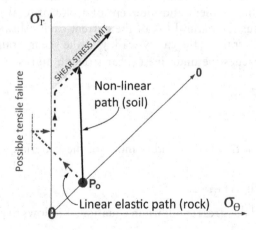

Figure A.21 Circumferential stress path.

A.4.4 Excess pore water pressure

Bolton and Whittle give a speculative description of non-linearity before yield for the undrained case where the mean normal stress increases during the initial loading but deformations occur at constant mean effective stress. Due to this increase in mean normal stress, there is some excess pore pressure, u, generated during the initial non-linear sub-yield loading:

$$\Delta \frac{1}{2}(\sigma_1 + \sigma_3) = (1 - \beta)\Delta p = u \qquad [a.38]$$

Post-yield, the development of excess pore pressure is obtained by deducting the total stress elements from [a.31]:

$$u = c_u[\ln(\gamma_c) - \ln(\gamma_y)] \qquad [a.39]$$

Combining the two and adding in the ambient pore water pressure, u_o:

$$u = u_o + c_u\left[\left(\frac{1 - \beta}{\beta}\right) + \ln(\gamma_c) - \ln(\gamma_y)\right] \qquad [a.40]$$

A.4.5 Conclusion for the undrained cavity expansion description

The definitive analysis for the undrained case is the Palmer numerical method, because of the minimal assumptions. The primary disadvantages are practical rather than theoretical. The solution is sensitive to the correct length of the strain scale. This demands an accurate assessment of the coordinate of radial stress and displacement at which the cavity expansion commences. Every time the direction of loading changes, a new coordinate at which the reversal commences needs to be decided, and subsequent stress-strain values are referred to this new origin. However, unlike the initial decision, identifying reversal points is generally straightforward.

The other disadvantage is the inherent numerical instability associated with the use of real data. This can be mitigated by curve fitting prior to starting the analysis and is the method adopted by Bolton and Whittle to derive non-linear parameters α and β from unload/reload cycles.

Assuming the shape of the stress-strain curve allows a closed-form solution to be applied, at least for the perfectly plastic case—there are as yet no such solutions for strain hardening or softening deformation. The non-linear elastic/perfectly plastic Bolton & Whittle (1999) analysis has been used to illustrate the closed-form approach. It is able to reproduce the non-linear behaviour prior to yield and recovers the linear elastic response as a particular case.

Deriving results is dependent on the match between the assumed shear stress-strain response and reality, but prior use of the Palmer analysis can indicate whether the correspondence is acceptable.

A.5 DRAINED CASE, FRICTIONAL MATERIAL

Material with a particle size greater than silt cannot under normal circumstances be sheared quickly enough to support the undrained condition. When a cavity is expanded in such material, no excess pore water pressure is generated. The stress path followed is drained and strength varies with changes to the normal stress (Fig. A.22).

Compared to the undrained path, additional arguments are needed to compute the development of strains and stresses. If following yield the shear stress/strain response is assumed to be perfectly plastic, then the shear stress changes at a constant stress ratio and at the peak angle of shearing resistance ϕ'. There are four additional unknowns that have to be decided:

- Ambient water pressure, u_0.
- Residual or constant volume friction angle, ϕ'_{cv}.
- Drained cohesion, c'
- Volumetric strains, ε_v

The ambient pore water pressure is required because the pressuremeter stress is total stress but the interpretation is implemented using effective stress. It is generally straightforward to identify u_0 from the closing stages of the pressuremeter field curve.

ϕ'_{cv} must be estimated or given—it is the ultimate shearing angle towards which the trend is aiming.

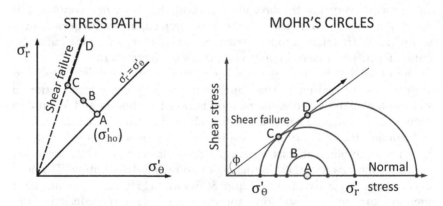

Figure A.22 Shearing in frictional material.

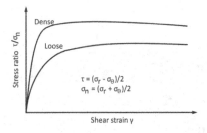

Figure A.23 Stress ratio/shear strain—actual.

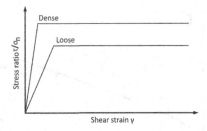

Figure A.24 Stress ratio/shear strain—ideal.

For the purely frictional case, c' is zero. Purely frictional solutions can be modified to include for drained cohesion but field data cannot be manipulated to find cohesion independently of the other parameters. Consequently, modelling the experimental data and adjusting σ'_{ho} and c' for the optimum fit is the only way to determine cohesion in the presence of friction.

The general form of the purely frictional plane strain shear test is shown in Fig. A.23 and the idealised response suggested by Hughes et al. (1977) is shown in Fig. A.24.

For the initial loading, shearing does not take place at constant volume. The change of volume in response to shearing is referred to as dilation. This can be positive or negative depending on whether the material is in a dense or loose condition at the start of shearing. Dilation causes volumetric strains, a relationship that is sketched in Fig. A.25 and Fig. A.26.

ψ is the angle of dilatancy. As the figures indicate, the actual response of the material at larger strain deformation is tending to a critical condition where it will deform at constant volume and the angle of shearing resistance becomes ϕ'_{cv}. Most closed-form solutions assume this occurs at strains significantly greater than those achieved in a normal test and use the idealised response shown in Fig. A.26.

Dilation can be explained using the "sawtooth blades" analogy (Newland & Allely, 1957). See Fig. A.27. As a unit of material is sheared,

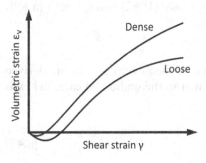

Figure A.25 Volumetric strains—ACTUAL.

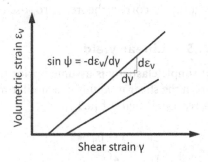

Figure A.26 Volumetric strains—IDEALISED.

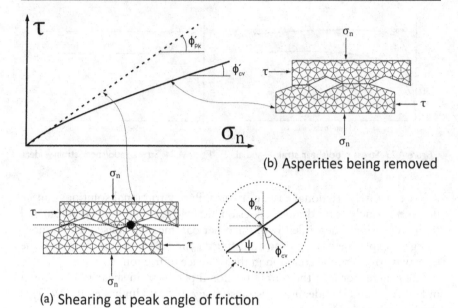

(b) Asperities being removed

(a) Shearing at peak angle of friction

Figure A.27 The sawtooth blades model (after Newland & Allely, '57).

particles ride over each other, changing the void space. With increasing normal stress, small asperities begin to be removed and breakage occurs, tending to a condition of zero dilation. Using this description, Bolton ('86) argues that the internal angle of shearing is essentially the sum of ϕ'_{cv} and ψ. He offers the following semi-empirical formulation for plane strain shearing:

$$\phi'_{pk} = \phi'_{cv} + 0.8\psi \qquad\qquad [a.41]$$

For all practical purposes, [a.27] is operationally indistinguishable from the rigorously correct theorem proposed by Rowe (1962) and given in [a.69].

A.5.1 Linear yield

If simple elasticity is assumed prior to yield, then cavity strains are derived from the shear modulus in a similar manner to the undrained case. Below is a rearrangement of [a.17]:

$$\varepsilon_c = \frac{\xi a}{r_a} = \frac{p - p_0}{2G} \qquad\qquad [a.42]$$

This relationship ceases at the yield stress p'_f given by:

$$p'_f = p'_o + p'_o \sin \phi' \qquad [a.43]$$

The cavity strain to reach p'_f is:

$$\varepsilon_{ip} = \frac{p'_o \sin \phi'}{2G} \qquad [a.44]$$

Once the pressure at the cavity wall exceeds p'_f, shearing continues at a constant stress ratio N where:[2]

$$N = \sigma'_r/\sigma'_\theta = \frac{1 + \sin \phi'}{1 - \sin \phi'} \qquad [a.45]$$

If this is combined with the stress equilibrium equation [a.7], then the stress state in the failed region at any radius can be calculated from:

$$\left(\frac{p'}{p'_f}\right)^{\frac{1+\sin\phi'}{2\sin\phi'}} = \frac{R}{a} \qquad [a.46]$$

R is the radius of the elastic/plastic boundary, a is the current cavity radius.

A.5.2 Non-linear yield

For the *undrained* analysis, it has been argued that the asymmetrical elastic response of the soil prior to yield can be described by a power law (Bolton & Whittle, 1999). The consequence is that the change of radial stress represented by $p'_f - p'_o$ is not equal to the shear stress at failure, τ, but τ/β, where β is the exponent of non-linearity. Applying this to the case of a drained expansion and assuming that whilst the shear stress is below the yield condition there are negligible volumetric strains, then the principal stresses at first yield can be written:

$$\sigma'_r = p'_f = p'_o + \frac{\tau}{\beta} \qquad [a.47]$$
$$\sigma'_\theta = \sigma'_r - 2\tau$$

For a perfectly frictional material, development of the plastic zone occurs at a constant stress ratio, with the radial stress being the major principal stress. At yield, [a.45] applies. Hence, combining this with [a.47]:

$$Np'_o + \frac{N}{\beta}\tau - 2N\tau = p'_o + \frac{\tau}{\beta} \qquad [a.48]$$

This allows τ/β to be derived:

$$\frac{\tau}{\beta} = \frac{p'_o(N-1)}{N(2\beta-1)+1} \qquad [a.49]$$

Substituting into [a.48]:

$$\sigma'_r = p'_f = p'_o + \left(\frac{p'_o(N-1)}{N(2\beta-1)+1}\right) \qquad [a.50]$$

Alternatively, using the friction angle ϕ' in place of N:

$$\sigma'_r = p'_f = p'_o + \left(\frac{p'_o \sin\phi'}{\beta(1+\sin\phi')-\sin\phi'}\right) \qquad [a.51]$$

The final unloading starts with the radial stress at a maximum p'_{mx} and "non-linear" yield first occurs at the borehole wall when the radial stress at the borehole wall is p'_{fu}:

$$\sigma'_r = p'_{fu} = p'_{mx} - \frac{2\tau}{\beta} \qquad [a.52]$$

The circumferential stress, σ'_θ, at contraction yield will be $\sigma'_r + 2\tau$:

$$\sigma'_\theta = p'_{mx} - \frac{2\tau}{\beta} + 2\tau \qquad [a.53]$$

The radial stress represented by $2\tau/\beta$ is discovered in a similar way to the elastic loading equations noting that yield in contraction occurs with the circumferential stress being the major principal stress; hence, $\sigma'_\theta/\sigma'_r = N$:

$$\frac{2\tau}{\beta} = \left(\frac{p'_{mx}(N-1)}{N-1+\beta}\right) \qquad [a.54]$$

The equivalent to [a.50] for the final unloading is:

$$\sigma'_r = p'_{fu} = p'_{mx} - \left(\frac{p'_{mx}(N-1)}{(N-1+\beta)}\right) \qquad [a.55]$$

Alternatively, in terms of $\sin \phi'$:

$$\sigma'_r = p'_{fu} = p'_{mx} - \left(\frac{2p'_{mx} \sin \phi'}{\beta(1 - \sin \phi') + 2 \sin \phi'} \right) \qquad [a.56]$$

For a $c' - $ phi material, the failure does not occur at a constant stress ratio, but can be made to seem so if all stresses are raised by $c' \cot \phi'$ (Carter et al., 1986).
  For the loading case:

$$\frac{\sigma'_r + c' \cot \phi'}{\sigma'_\theta + c' \cot \phi'} = N = \frac{p'_o + \frac{\tau}{\beta} + c' \cot \phi'}{p'_o + \frac{\tau}{\beta} - 2\tau + c' \cot \phi'} \qquad [a.57]$$

This leads to:

$$\sigma'_r = p'_f = p'_o + \left(\frac{(p'_o + c' \cot \phi')(N - 1)}{N(2\beta - 1) + 1} \right) \qquad [a.58]$$

Alternatively, in terms of ϕ':

$$\sigma'_r = p'_f = p'_o + \left(\frac{(p'_o \sin \phi' + c' \cos \phi')}{\beta(1 + \sin \phi') - \sin \phi'} \right) \qquad [a.59]$$

Similarly, the expression for first yield in unloading in a c'-phi material is obtained by raising the principal stresses by $c' \cot \phi'$:
  At yield in contraction:

$$\frac{\sigma'_\theta + c' \cot \phi'}{\sigma'_r + c' \cot \phi'} = N = \frac{p'_{mx} - \frac{2\tau}{\beta} + 2\tau + c' \cot \phi'}{p'_{mx} - \frac{2\tau}{\beta} + c' \cot \phi'} \qquad [a.60]$$

This leads to the following for the yielding stress in unloading:

$$\sigma'_r = p'_{fu} = p'_{mx} - \left(\frac{(p'_{mx} - c' \cot \phi')(N - 1)}{(N - 1 + \beta)} \right) \qquad [a.61]$$

Alternatively:

$$\sigma'_r = p'_{fu} = p'_{mx} - \left(\frac{2(p'_{mx} \sin \phi' - c' \cos \phi')}{\beta(1 - \sin \phi') + 2 \sin \phi'} \right) \qquad [a.62]$$

If $\beta = 1$, the value for linear elasticity, all non-linear equations revert to Mohr-Coulomb failure definitions. Typical values for β in granular deposits are 0.7–0.85, the lower value tending to be the characteristic of a more fine-grained material.

Given these non-linear definitions of p'_f or p'_{fu}, non-linear strain at yield ε_{ip} and ε_{ipu} can be derived as follows, based on values for α and β obtained from unload/reload cycles:

$$\varepsilon_{ip} = \left(\frac{p'_f - p'_o}{a_r} \right)^{\frac{1}{\beta}} \qquad\qquad [a.63]$$

$$\varepsilon_{ipu} = \left(\frac{p'_{mx} - p'_{fu}}{\alpha_{ru}} \right)^{\frac{1}{\beta}} \qquad\qquad [a.64]$$

It is likely that, due to changes in mean effective stress, the shear modulus at first yield will be different from that of the final yield in contraction. Additionally, a_r and α_{ru} are obtained from the derived α values to give the non-linear response in radial stress–cavity strain space:

$$\alpha_r = 10^{\left(\log \frac{\alpha}{\beta} + \beta \log 2 \right)} \qquad\qquad [a.65]$$

A.5.3 Drained yielding

The Hughes et al. solution assumes a perfectly plastic response once the material has failed. It is combined with the sub-yield definitions to give the following expression for cavity strain:

$$\varepsilon_c = \varepsilon_{ip}(1 + \sin\psi)\left[\left(\frac{p'}{p'_f} \right)^{\frac{1}{s}} - \left(\frac{\sin\psi}{(1 + \sin\psi)} \right) \right] \qquad\qquad [a.66]$$

The S in the exponent term is defined as:

$$S = \frac{(1 + \sin\psi)\sin\phi'}{1 + \sin\phi'} \qquad\qquad [a.67]$$

Based on Carter et al. (1986), [a.66] can be adapted to include for drained cohesion:

$$\varepsilon_c = \varepsilon_{ip}(1 + \sin\psi)\left[\left(\frac{p'_c + c'\cot\phi}{p'_f + c'\cot\phi} \right)^{\frac{1}{s}} - \left(\frac{\sin\psi}{(1 + \sin\psi)} \right) \right] \qquad\qquad [a.68]$$

In either version of the solution, the yield coordinate ε_{ip} and p'_f can be quantified using linear or non-linear definitions.

An inspection of [a.66] shows it is similar to the basic stress formulation given by [a.46]. The difference in the exponent arises from the introduction of Rowe's dilatancy rule into the solution. This can be written as:

$$\frac{(1 + \sin\phi'_{pk})}{(1 - \sin\phi'_{pk})} = \frac{(1 + \sin\phi'_{cv})}{(1 - \sin\phi'_{cv})}\frac{(1 + \sin\psi)}{(1 - \sin\psi)} \qquad [a.69]$$

For the purely frictional case, the S in the power term can be obtained by plotting measured values for effective stress and cavity strain on log scales and finding the ultimate gradient, as in Fig. A.28. Equ. [a.67] and [a.69] together then allow either ϕ' or ψ to be obtained (Fig. A.29).

The assumption of perfect plasticity means the analysis is generally used to derive a single value for ϕ'. This is an imposed constraint for the closed form solution; it is not a fundamental requirement. Manassero (1989) points out that Rowe's dilatancy relationship is applicable for all elements on the stress-strain path, not just at peak strength. From this it follows that the local gradient of the field curve in log-log space combined with [a.69] provides the current friction angle. Given ϕ' for all points on the loading curve, [a.45] allows the stress ratio to be calculated. it is then possible to extract several important relationships from the experimental data.

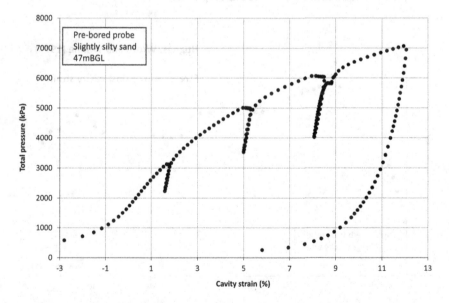

Figure A.28 Pre-bored test in sand.

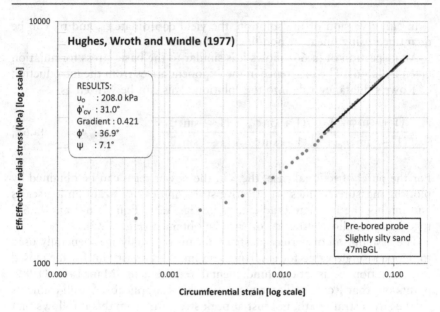

Figure A.29 Hughes, Wroth & Windle (1977).

Fig. A.30 to Fig. A.33 give an indication of what is possible, whilst recognising that numerical instability inevitably introduces false peaks and troughs in what should be a smooth response.

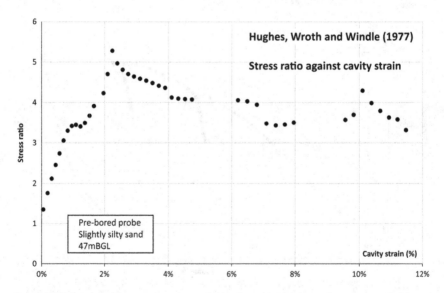

Figure A.30 Stress ratio.

As indicated in Fig. A.31 and Fig. A.33, towards the end of the loading, dilation has significantly reduced and the material is starting to deform without generating further volumetric strains. This makes it possible to calculate the effective limit pressure for indefinite expansion.

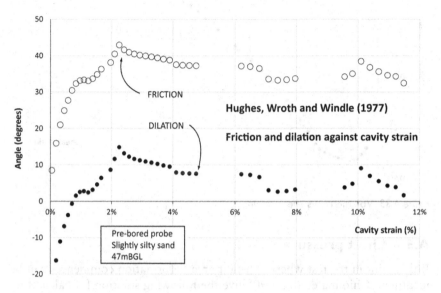

Figure A.31 Friction and dilation angles.

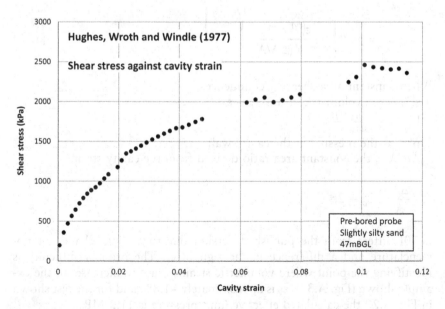

Figure A.32 Shear stress.

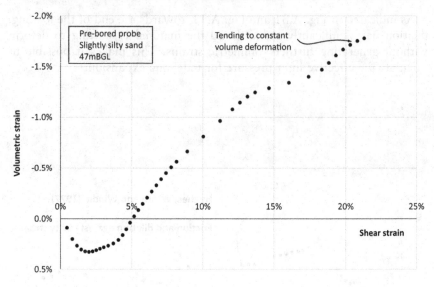

Figure A.33 Volumetric vs shear strain.

A.5.4 Limit pressure

If the point in the test where constant area deformation commences can be identified, Ghionna et al. (1990) give the following solution for calculating the effective limit pressure:

$$
p'_{lim} = p'^{(cv)} \left[\frac{\varepsilon_v^{(cv)} + 1}{\varepsilon_v^{(cv)} + (\Delta A/A)} \right]^{\left(\frac{1 - k_a^{cv}}{2} \right)} \tag{a.70}
$$

When constant area shearing commences:
$\varepsilon_v^{(cv)}$ is the volumetric strain, $\varepsilon_r - \varepsilon_c$
k_a^{cv} is $\frac{(1 - \sin\phi'_{cv})}{(1 + \sin\phi'_{cv})}$
$p'^{(cv)}$ is the pressure at the cavity wall
$\Delta A/A$ is the constant area ratio derived from the cavity strain:

$$
\frac{\Delta A}{A} = \varepsilon_c^2 - 2\varepsilon_c
$$

[a.70] differs from the published version due to a slight change of nomenclature and a difference in the sign of ε_c. The primary difficulty is identifying the point where volumetric strains cease to increase. In the example shown (Fig. A.33), ε_v is approximately –1.8% and for the test shown in Fig. A.28 the calculated effective limit pressure is 11.2 MPa.

A.5.5 Conclusion for the drained cavity expansion description

The core analysis for the drained case is the Hughes et al., (1977) solution. Although usually applied as a closed-form solution, it can also be implemented as a numerical method. All other solutions such as Carter et al. (1986) are developments of the Hughes et al. analysis.

The advantage of the closed-form approach is the ability to construct a model that can be used to compare experimental data to the interpreted parameter set. For this to occur with appropriate constraints, the cavity contraction has also to be included. Although solutions for the contraction are available, frictional material will only contract at peak strength for a very limited strain range. All modelling based on closed-form solutions will therefore be ambiguous compared to the constrained undrained models.

Modelling the drained test with contraction included is presented in Chapter 5.

NOTES

1 Also independently by Baguelin et al. (1972) and Ladanyi (1972).
2 Hughes et al. (1977) define N as the reciprocal of this. Most subsequent papers give N as major/minor stress and we have followed this convention.

REFERENCES

Baguelin, F., Jezequel, J., Le Mee, E. & Le Mehaute, A. (1972) Expansion of a cylindrical probe in cohesive soils. *Journal of the Soil Mechanics and Foundations Division, ASCE* 98 (SM11), Paper 9377, pp. 1129–1142.

Bolton, M.D. (1986) The strength and dilatancy of sands. *Géotechnique* 36 (1), pp. 65–78.

Bolton, M.D. & Whittle, R.W. (1999) A non-linear elastic/perfectly plastic analysis for plane strain undrained expansion tests. *Géotechnique* 49 (1), pp. 133–141.

Carter, J.P., Booker, J.R. & Yeung S.K. (1986) Cavity expansion in cohesive frictional soil. *Géotechnique* 36 (3), pp. 349–358.

Ghionna, V.N., Jamiolkowski, M., & Manassero, M. (1990). Limit pressure in expansion of cylindrical cavity in sand. *Pressuremeters. Third International symposium*, London: Thomas Telford, pp. 149–158.

Gibson, R.E. & Anderson, W.F. (1961) In situ measurement of soil properties with the pressuremeter. *Civil Engineering and Public Works Review* 56 (No. 658 May), pp. 615–618.

Hughes, J.M.O., Wroth, C.P. & Windle, D. (1977) Pressuremeter tests in sands. *Géotechnique* 27 (4), pp. 455–477.

Ladanyi, B. (1972) In situ determination of undrained stress-strain behaviour of sensitive clays with the pressuremeter. *Canadian Geotechnical* 9, pp. 313–319.

Manassero, M. (1989) Stress-strain relationships from drained self-boring pressuremeter tests in sands. *Géotechnique* **39** (2), pp. 293–307.

Newland, P.L. & Allely, B.H. (1957) Volume changes in drained triaxial tests on granular material. *Géotechnique* **7** (1), pp. 17–34.

Palmer, A.C. (1972) Undrained plane-strain expansion of a cylindrical cavity in clay: a simple interpretation of the pressuremeter test. *Géotechnique* **22** (3), pp. 451–457.

Rowe, P.W. (1962) The stress dilatancy relation for static equilibrium of an assembly of particles in *Contact. Proceedings of the Royal Society* **269** (Series A), pp. 500–527.

Timoshenko, S.P. & Goodier, J.N. (1934) *Theory of elasticity.* McGraw Hill.

Wood, D.M. & Wroth, C.P. (1977) Some laboratory experiments relating to the results of pressuremeter tests. *Géotechnique* **27** (2), pp. 81–201.

Technique, Equipment and Procedure

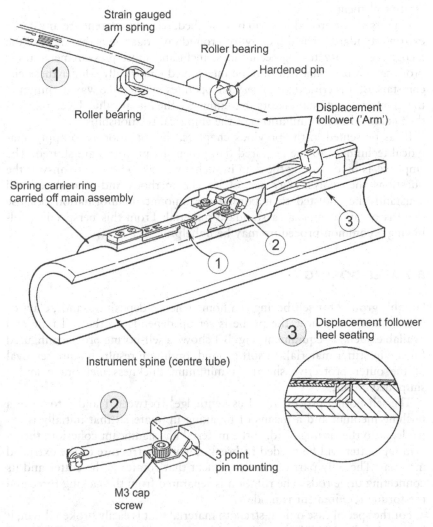

Strain gauged
arm spring

Roller bearing

Hardened pin

Roller bearing

Displacement
follower ('Arm')

Spring carrier ring
carried off main assembly

Displacement follower
heel seating

Instrument spine (centre tube)

3 point
pin mounting

M3 cap
screw

Figure B.0 Details of self-boring pressuremeter displacement measuring system.

B.I INTRODUCTION

This appendix describes three of the commercially available devices whose data have been used elsewhere in this text. The advantages and limitations of the three deployment systems (self-boring, pre-boring and pushing) are developed in sufficient detail to aid an engineer commissioning tests in making appropriate choices.

The primary variable is the material but for all three probes described there is an overlap, so additional factors come into consideration. For almost all pressuremeter testing, some support from standard drilling operations will be required. Ultimately, the cost of these additional components will be the limiting element.

Aspects of test procedures will be described, without reference to any of the existing standards. The difficulty for ground investigation is that most of the recognised *in situ* testing techniques, including most pressuremeter tests, produce numbers that can only be interpreted empirically. Under these circumstances, it is critical that everyone operates in the same way to minimise the number of external factors that influence the initial value. Uncertainty is then confined to the limitations of the empirical relationships.

If, as presented in the previous chapters, the intention is to apply analytical techniques to the acquired data, then the priorities are shifted. The emphasis is on obtaining the data in such a way as to be able to answer the questions the analyst may ask of it. The methods and procedures will constantly be adjusted to maximise the amount of information because there is no such thing as a standard material. From this perspective, following a common procedure may be unhelpful.

B.2 SELF-BORING

Suitable ground for self-boring is a homogeneous deposit of sands, clays or soft rock. Exactly how the probe is set up depends on the soil type and available ancillary equipment. Fig. B.1 shows a self-boring probe configured for use in stiffer material. In soft ground, optimum results require removal of the outer protective sheath, a minimum thickness membrane and a simplified cutter.

The self-boring process requires kentledge[1] (between 1 and 4 tonnes), a flushing medium and a means of breaking up material that initially is extruded into the cutting head. If the material has significant cohesion, then a rotating cutter will be needed to break down the structure of the extruded material. The only part of the instrument that rotates is the cutter and its connecting drive rods. The rotation is separated from the jacking force and the torque requirement is modest.

For the special case of low-strength material that is easily broken down, it is occasionally possible to replace the rotating cutter with a jetting lance

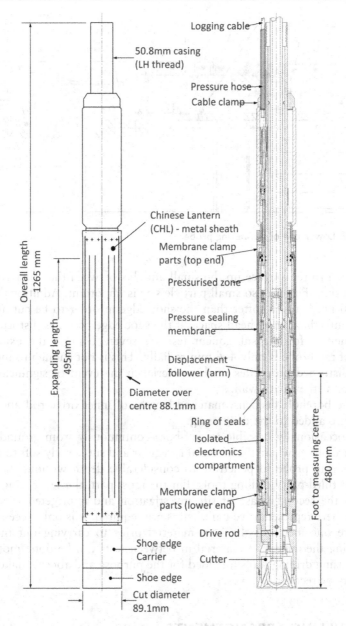

Figure B.1 Self-boring pressuremeter (strong ground version).

(see Chapter 5). The advantage of this method is the absence of a need for rotational force and hence a greatly simplified insertion system.

The self-boring pressuremeter cannot be used in material such as gravel. The difficulty is twofold. All the material that enters the cutting head has to

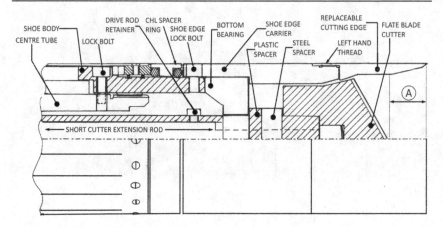

Figure B.2 Lower end of self-boring probe.

be flushed to the surface up the small annulus between the inner rod and outer casing (Fig. B.2) so small particle size is important. Additionally, the material needs to be softer than the shoe edge in order to be cut and extruded into the cutting head space. If the shoe edge becomes distorted, the consequences for the subsequent test are severe because the expansion range of the probe is only 4–6 mm radially. This is not enough to mobilise the undisturbed properties of the material in the event of significant disturbance at the cavity wall.

As the borehole deepens, matched lengths of inner drive rod and outer casing are added to the drill string.

It is occasionally possible to self-bore continuously from ground level. There are places where the material is regular and sufficiently soft to permit a self-boring probe to penetrate to considerable depth without the assistance of third-party drilling tools. For the most part, the self-boring test is used in the second phase of site investigation and is targeted at specific layers where representative parameters are required. It is only necessary to self-bore one metre of the host material prior to carrying out the test. Removing the metres of material in between tests can be done more efficiently using drilling tools intended for the purpose and able to make other soil tests or recover samples.

B.3 DRILLING ARRANGEMENTS

B.3.1 Self-boring with a cable percussion rig

A typical arrangement, widely used, is shown in Fig. B.3. This shows a self-contained and portable drilling system working in conjunction with a cable percussion (CP) rig.

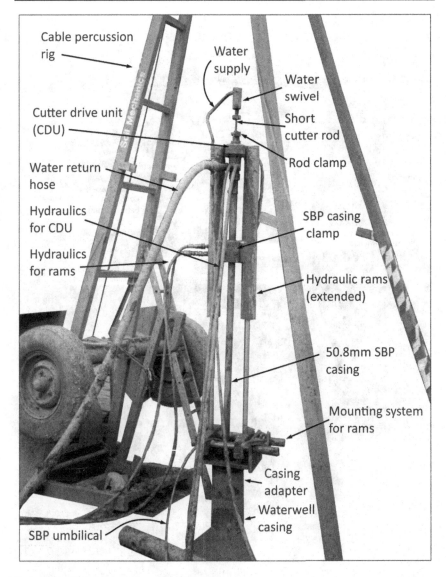

Figure B.3 SBP with a cable percussion rig.

The system consists of:

- Hydraulic rams to provide down thrust
- A motor to rotate the cutter drive rods
- A water pump with a hydraulic drive
- An engine-driven hydraulic pump and suitable distribution controls for the above parts (see Fig. B.4).

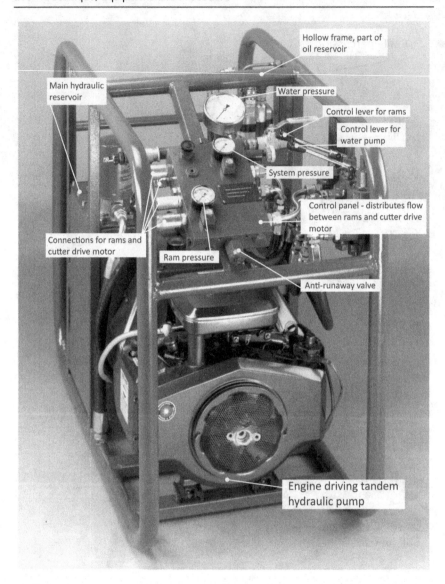

Figure B.4 Portable power pack.

- A connection to a column of water well casing that has been hammered into the ground. It is the skin friction on the casing that provides kentledge for the rams.

Although the CP system is generally deprecated, it still has a role because of its versatility. Cut-down versions (electrically powered) can operate inside buildings inaccessible to more complex drilling machines. They are excellent tools for making a cased hole through gravels.

A problem with the CP operation is the self-boring system requirement for a water flush. Cable percussion work is carried out dry, but the self-boring system requires water. This is recycled but will still need replacing after a few metres of boring because it becomes too thick with cuttings to be effective. In addition, whenever the SBP and its rods are removed from the borehole some water gets left behind and the borehole will require bailing.

The disadvantages of the combined cable percussion and self-boring operation are of two kinds. If the material is clay, the cased length of the borehole will be a few metres only and in general the borehole will be expected to stay open. As explained in the cavity expansion experiments in Chapter 8, where a pressuremeter test has been carried out, the radius of the material that has gone plastic is probably 10 times that of the pressuremeter itself. These plastic episodes are potential zones of borehole collapse. Whilst the drilling tools are being regularly deployed, the collapse will be deferred. However, a period of standing (perhaps over a weekend) may result in a borehole collapse from a level several metres above the penetrated depth.

The second problem is granular material. These layers will not stand open and the cable percussion system will have to case the entire borehole. The casing is always surged into place and installed in this way cannot provide sufficient kentledge (Fig. B.5).

Figure B.5 Cuttings recovered from return flush.

Provided that there is a noticeable clay layer closer to the surface, then a length of larger diameter casing can be installed. It must be hammered in for the last 0.5 metres. Thereafter, it is left undisturbed. It is to this casing that the SBP system will connect. If it is found that the entire borehole needs casing it will be done at a smaller size that can be loosely placed—the SBP system will continue to connect to the larger-diameter casing.

The maximum length of the twin drill string that can be added with a standard cable percussion rig is 3 metres at any one time. Experience suggests that 30 metres depth is the economic limit for this method of working.

B.3.2 Self-boring with a rotary drill rig

Fig. B.6 is a sketch of the kit required to operate the self-boring probe from the drive head of a standard rotary drill rig. The technical problem is separating

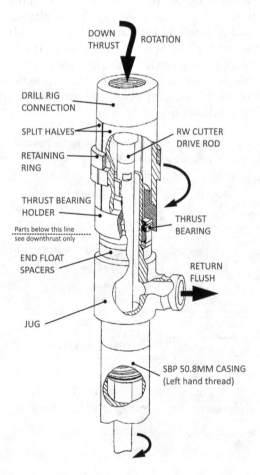

Figure B.6 Rotary rig adapter for SBP.

down thrust from rotation so that each part of the two rod system sees the appropriate force. This is achieved by placing a large thrust bearing between the rig drive and the outer casing of the SBP. All other parts are concerned with ensuring access to the inner drive rod when required and setting the end-float. The drive rod must spin freely and be length adjusted so that the cutter is in the correct position inside the cutting head.

The system is versatile and can be made to work with almost any rotary drive system. Kentledge is obtained from the self-weight of the drill rig, and water flush is also provided by the rig.

A drill rig suitable for this way of working needs to weigh at least 4 tonnes. This is not a significant problem for rigs operating in heavily over-consolidated material typical of North America and Europe, but in other parts of the world smaller rigs are often the only available option. One way to manage this weight limitation is to reduce the kentledge requirement by setting up the SBP for strong ground work (Fig. B.1) even though the material is comparatively soft. The reduced skin friction permits the SBP to be used with modest downthrust at the expense of a small amount of stress relief.

The system also requires at least 1.5 metres of clearance between the rotary rig drive head when fully retracted and the top of the cased hole. See Fig. B.7. This shows a typical rotary drill rig in operation with the adapter hardware clearly visible. One metre of the clearance is required for the SBP drill string, the remaining 0.5 metres for the adapter hardware and access to joints. If the clearance cannot be provided, as occasionally happens with smaller rigs, the only alternative is shorter lengths of drill string and this slows the drilling operation considerably.

All flush returns to the surface up the annulus of the instrument and is recovered via the ejection port mounted on the jug. The returns can be sieved for identification purposes and Fig. B.5 is one such example, from self-boring carried out in Boston Blue Clay. The material was broken down with a flat-blade cutter and has been recovered as flakes.

For almost all self-boring, the flushing medium is water. The boring process is greatly improved if drilling muds are added to the water supply. The standard two-part drill string is dimensioned for continuous boring, noting that twice the diameter of the umbilical hose has to be taken into account. This means that the downpipe has too narrow a bore and return path annulus to cope with a pure air flush. Non-standard arrangements of drill string sizes have been used with air (see Chapter 8) but these are specifically for "one test and out" procedures—the necessarily large diameter of the outer casing does not permit self-boring for more than the length of the instrument.

B.4 DRILLING VARIABLES

Previous references to self-boring have stressed the importance of the relationship of the cutting blade to the shoe edge (marked "A" in Fig. B.2).

Figure B.7 SBP operating under a rotary drill rig.

This defines the amount of material allowed to enter the cutting head space before the rotating cutter can break it down. Stiff material will have a small gap, loose or soft material a larger gap.

In practice it is not as critical as the literature suggests. For any given cutting gap, there is a rate of advance and flushing fluid volume that is ideal. Generally self-boring is like rotary coring—if the full flush is returning though the instrument annulus and the instrument is advancing then there is little to be concerned about. The efficiency of the cutting is the most important element—the cutting head must not become blocked. If this occurs in a clay, the SBP piezometers will instantly register a sharp rise in excess pore water pressure with consequences for the subsequent test, even if the blockage is cleared.

Cutters can take many forms, from a simple flat blade to a rock roller or full face cutter. This is used when self-boring into soft rock formations and requires a modified cutting head as the cutter is now seeing some torque.

Most SBP testing is carried out using a small drag bit.

B.5 SELF-BORING TEST PROCEDURES

B.5.1 Undrained

Whilst the SBP is drilling, the instrument should be "live" and data recorded. Especially in a cohesive material, the trend of pore water pressure against time can be informative.

The drilling-in of the probe and the subsequent test are a continuous process and there needs to be the minimum of delay between completing one and starting the other. If disturbance has created excess pore water pressure, then pausing after drilling allows the material to consolidate and alters the *in situ* properties, in particular strength. If the material has been stress relieved then the sooner a load is applied to the cavity wall the better, to limit the opportunities for slumping. In neither case will waiting result in the material becoming "undisturbed."

A standard undrained self-boring test that can be used as a template is shown in Fig. B.8.

The important points are these:

1. This is an undrained test. The rate at which the cavity expands has to be fast enough to prevent the loss of excess pore water pressure. The minimum permissible rate will vary depending on the permeability of the material.

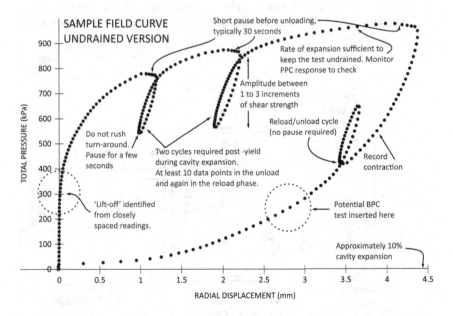

Figure B.8 SBP test example, undrained.

2. There are places in the test where events must be well-defined by many quiet data points and places where the form of the test is sufficiently represented by more widely spaced data.

3. Although it is important to have at least two credible sources of stiffness data from the expansion phase of the test, each cycle is an interruption that will tend to encourage drainage. Once this occurs the material properties start to change so the number of unload/reload events should be no more than necessary to preserve the integrity of the test as a whole.

4. The full cavity contraction must be recorded.

5. Reload/unload data have the same validity as unload/reload data and fewer consequences for the material. However deferring the taking of cycles solely to the cavity contraction is a high-risk strategy—the membrane may rupture without warning and the cavity contraction will be lost.

6. The instrument is inflated using compressed air or nitrogen. The test can be run in a semi-automatic mode using a strain control unit (Fig. B.9) to control the application of pressure and keep the rate of expansion constant. This would be relevant if all elements in the soil mass saw the same strain (as is the case for a triaxial test), but this is not true for a cavity expansion test.

   It happens that for most of the test it is inappropriate to force the material to respond at a constant strain rate. Whilst the material response is elastic or pseudo-elastic and hence the accumulation of strain is small, the load application should be done at a constant rate

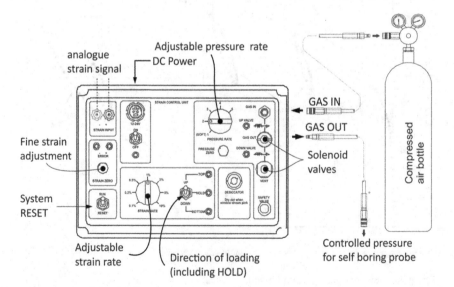

Figure B.9 Schematic of a strain control system.

of stress. It is only the post-yield response where an argument can be made for constant strain.

The automatic control system also introduces small pressure pulses as valves open and close and can make the data path seem noisy when examined very closely. An operator is able to achieve a smoother loading using a high-quality manual system (see Fig. B.16). This is how the test in the example was carried out.

However, an automatic system is essential for carrying out consolidation testing where the size of the cavity is kept constant whilst the excess pore pressures are allowed to dissipate.

7. The test example does not include a Balance Pressure Check procedure, but the place where this could be carried out is indicated. It covers the pressure range that includes the "lift-off" stress from the initial part of the test.

B.5.2 Consolidation testing

The recommended clay test procedure requires slight modification for a consolidation or holding test. The test priority is preservation of the excess pore water pressure generated by a cavity expansion. Every interruption to the loading, such as taking an unload/reload cycle, is potentially a point where drainage can be initiated. One good cycle is essential for the analysis; any more is an unnecessary risk.

The consolidation data described in Chapter 6 were obtained with the use of a semi-automatic control system (Fig. B.9). This gives a constant strain hold and permits the use of relatively straightforward analytical solutions.

B.5.3 Drained

Fig. B.10 is a template for a test in sand.

1. The main difference between the drained and undrained test is the additional unload/reload cycles during the expansion phase. Modulus varies with the mean effective stress so adequately evaluating this response requires a minimum of four cycles whilst the cavity is expanding.
2. In some siltier deposits, partial excess pore pressure may develop if the expansion is carried out too quickly, so the rate of advance should be adjusted to ensure this does not occur. Time-dependent effects when changing the direction of loading are less of a difficulty for the drained test due to the absence of excess water pore pressures.
3. Towards the end of the cavity contraction phase, the membrane will lose contact with the cavity wall. This provides useful data for the ambient pore water pressure and should not, therefore, be rushed.

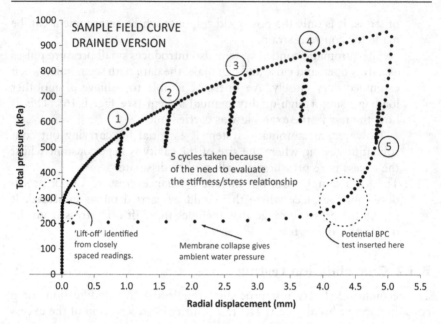

Figure B.10 SBP test example, drained.

B.5.4 Summary of requirements and considerations

- Is the material to be tested reasonably homogeneous and free of hard particles greater in size than 1 mm?
- Can self-boring be carried out without the assistance of a drill rig?
- If a drill rig is required, will it be a rotary or static type?
- Will there be sufficient weight for the self boring system to react against?
- Is there an adequate supply of clean water?
- Is there a method of disposing of the slurry created by the self-boring system?
- What is the maximum length of drill string that can be added at any one time?
- If using a rotary rig, what is the available clearance between the drill head and the operations level?
- When using a rotary rig, does it have an adequate water pump and delivery system?

These are in addition to the usual site working logistical issues that are a part of every ground investigation.

B.6 PRE-BORED TESTING

Pre-boring is the preferred option when the material is mixed (boulder clay, flint-laden chalk) or expected to fail in tension (rock). It is also the option

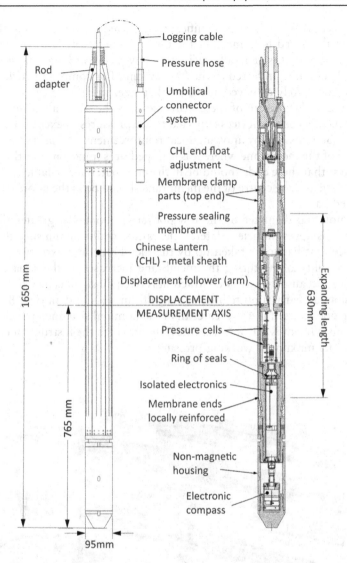

Figure B.11 Schematic of 95 mm HPD.

for tests that require pressures greater than 10 MPa to develop a full cavity expansion. This might be a deep deposit of very dense sand.

The test cavity is pre-formed using standard drilling tools. The instrument itself is dimensioned to fit inside a pocket formed by a standard core barrel — the 95 mm high-pressure dilatometer (HPD, Fig. B.11) is intended for an "H" size cavity (approximately 101 mm). There is also a version of the instrument for operating in an "N" size pocket (approximately 76 mm). As can be seen from the dimensions, it is possible to make a test in a 1.5 metre long pocket.

The success of the test is significantly affected by how closely the diameter of the cored hole matches the outer diameter of the coring tool. In poorer material, it is generally easier to recover core at larger diameters, so the "H" system is preferred to the "N" system. For competent rock, coring may be adequately achieved at the smaller size.

The particular feature of these instruments is the unusually thin pressure sealing membrane for devices expected to see pressures in excess of 20 MPa. This is made possible by applying local reinforcement (extrusion limiters) at the ends of the membrane where it is clamped. It is important for the success of the test that these ends should be in the best material available for a given test section, and the centre of the membrane should be in the poorest or most fractured part.

The nominal diameter of the test pocket is about 4% greater than the probe. This gives sufficient lateral clearance, while minimising the axial clearance that has to be taken up by the membrane and extrusion limiters. The possibility of rupturing the membrane increases as the square of the clearance. Because the cavity contraction is of equal significance to the cavity expansion, an early rupture implies a significant loss of data (Fig. B.12).

If the pocket itself is 4% greater than the nominal size, then the clearance for the instrument is 10%. This is the limit at which the instrument could be taken to its maximum working pressure.

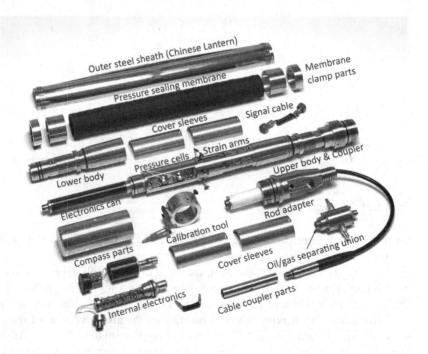

Figure B.12 95 mm high-pressure dilatometer and parts.

Unlike the self-boring probe built upon a hollow tube, the displacement measurement system of the pre-bored tools can exploit the full diameter and can therefore go to a much greater expansion. This permits it to make a complete loading curve to significant strain even if the pocket is a little oversize.

B.6.1 Complex hole preparation

In difficult materials, elaborate strategies may be required in order to obtain a pocket that can be tested by pre-bored means.

One such is pressuremeter testing in material such as limestone with solution cavities. The ground is for the most part competent and may even be cored reasonably successfully. However, the presence of the cavities (which may be any size from a coin to a small cavern) prevent the pressuremeter applying significant load to the cavity wall. An attempt to do so is likely to end with premature membrane loss at pressures too low to be useful in determining stiffness or strength.

One approach that has been used successfully several times is to start by coring a testing borehole to its full depth at the required size for the pressuremeter. It is then filled with a weak grout cement mixture and the amount of grout necessary to achieve this is recorded. This gives an indication of the extent of the cavities.

The grout is allowed to set, after which the borehole may be cored once more to its full depth. The cement cores are retained to prove cavity wall integrity and that the re-coring has not strayed off-axis and cut fresh material.

The hole is then tested in one sequence. The pressuremeter is placed at the bottom of the hole and after each test is pulled up to the next required level.

This approach requires a small compromise, in that it is always preferable to test the material as soon after coring a test pocket as can be managed.

A more complex version of this approach has been used to test cemented boulder clays, the remnants of ancient river systems severely consolidated from cycles of glaciation. It is a rock from which core cannot be obtained and test pockets cannot under normal circumstances be established (Fig. B.13).

The following method has been used with a 73 mm diameter pressuremeter — refer to Fig. B.14:

1. A 5-inch/125 mm diameter borehole is drilled with a down-hole air hammer and the drill casing is left in place to support the borehole.
2. A 1.95-inch/38 mm OD flush jointed steel pipe is set in the centre of the main casing and kept concentric using plastic centralisers.

Figure B.13 Dense gravel.   *Figure B.14* A cement sleeve.

3. This is used as a tremie pipe to fill the borehole with grout. The casing is unscrewed from the drill bit shoe and is slowly pulled up the hole whilst the grout is being injected.
4. Two weeks are required for the grout to fully harden. The central rod is over-cored with an N barrel and removed.
5. What remains is a cement sleeve, 19 mm wall thickness, where there had been casing. The pressuremeter is placed at the bottom of the hole and a sequence of tests is carried out, proceeding from the deepest to the shallowest.

The tests themselves are indistinguishable from standard tests because the cement sleeve adds very little to the strength and stiffness of the material. The calculation of cavity strain requires adjustment because of the additional thickness of the cement contribution to the pressuremeter membrane. Fig. B.15 is an example of what the cement sleeve itself looks like.

Procedures such as this seem expensive—each hole sacrifices the drill bit shoe. However, this may be the only way of obtaining sensible engineering data from a complex formation.

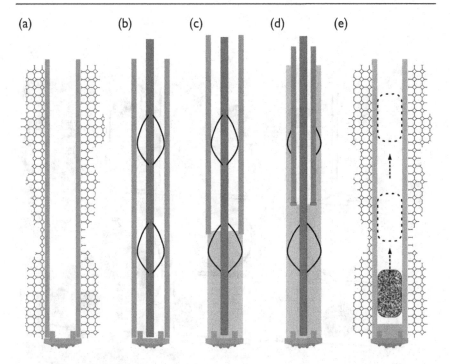

(a)   (b)   (c)   (d)   (e)

Figure B.15 Forming a cement sleeve.

B.6.2 Expansion method

The instruments can be inflated by oil or gas. Ideally, oil should be used on safety considerations. The stored energy is small and a burst carries minimal risk. In normal operations the oil is recovered and can be reused for subsequent tests. It is usually applied by a hand pump, so comparing the output from the sensors in the probe to the injected volume permits the operator to recognise when the membrane is expanding irregularly.

The preference for oil is complicated by two difficulties:

1. In a deep or dry hole, the head of liquid in the umbilical may be too great for the oil to return by itself and the probe will not deflate following a test.
2. The use of oil, even a bio-degradable one, is deprecated in environmentally sensitive areas.

The first of these objections can be overcome by connecting an additional pressure line to the probe. Unlike the umbilical, where the internal volume is largely occupied by a logging cable, the hydraulic resistance of the additional empty line is low. Fluid will return relatively freely. If the return

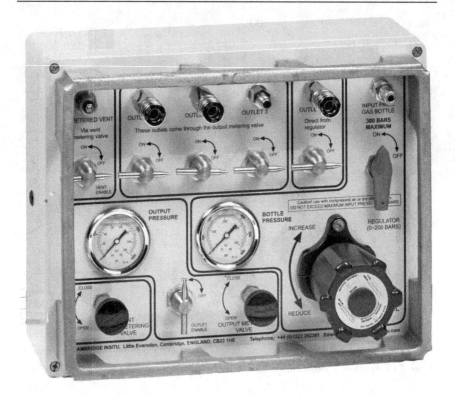

Figure B.16 Pneumatic control unit.

time is still too long, then air can be used to blow the oil back until the stiffness of the membrane is sufficient to close it onto the body of the probe. The penalty is the burden of handling an extra pressure line.

Alternatively, compressed air can be used to pressurise the probe. Pneumatic inflation is convenient, efficient and clean. Pressure is applied through a system of valves that permit a high degree of control over the stress application (Fig. B.16). The handling of pressurised gas bottles and parts demands extra care be taken. For maximum convenience and portability, conventional aqualung cylinders are used to store the air. These are kept full using a small commercially available compressor used to recharge diving bottles.

Apart from safety issues, the only significant disadvantage to the use of compressed air is greater sensitivity to temperature change. If the temperature at depth is very different from that where the air is stored at the surface then this will affect the transducer readings, especially when the direction of loading is reversed.

Fig. B.17 is an extreme example. This shows a pair of HPD tests carried out in the same pocket, at the same level, with the same instrument, on the same day. For the first test, air inflation was used, for the

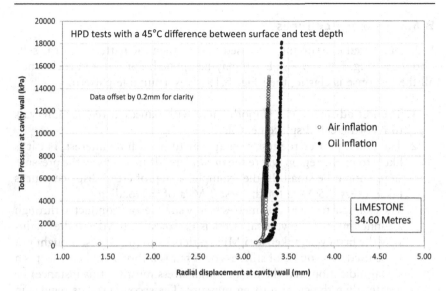

Figure B.17 Air and oil inflation compared (a).

second the probe was inflated with oil. At the scale of this figure, the two trends seem broadly similar, with a slight indication that the air inflation is noisier. It is only when the displacement scale is greatly expanded (Fig. B.18) does it become clear that the data from the air inflated test are unusable. It indicates the attention to detail required in order to obtain the best possible results.

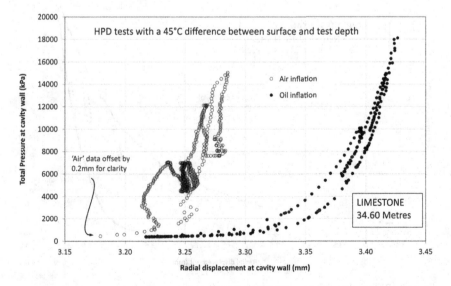

Figure B.18 Air and oil inflation compared (b).

B.6.3 Test procedures

If the pre-bored pressuremeter is used in soils then the patterns for the self-bored test (Fig. B.8 and Fig. B.10) may be applied. For a test in rock when all the response is elastic, then Fig. B.19 gives a suitable pattern.

1. The test advances as a smooth curve with pauses at regular intervals to record creep displacement data.
2. The initial part of the test may appear to be of little interest, but it is likely to be the region where the *in situ* lateral stress acts so the pressure steps should be small. In the example, a step of creep displacement is taken every 0.5 MPa for the first 3 MPa of the loading.
3. The example test was 300 metres below surface and conducted through a tunnel invert. The vertical stress is in excess of 6 MPa and explains why the first cycle, taken at 5 MPa, differs from the others. It ought not to be used as a source of stiffness data but is informative because it is a visual indication that the major initial stress must (in this instance) be greater than the cycle starting pressure. The second cycle is regular in form and consistent with cycles 3 and 4, so it may be inferred that the base of the second cycle is above the insitu lateral stress.
4. Additional cycles are taken at 10 MPa and 15 MPa to determine the repeatability of the modulus measurements. In an elastic material, the absolute amplitude of the cycles is not critical but the base of the cycle should not be allowed to go below the likely overburden stress. The cycle should be a well-defined smooth event and this requires

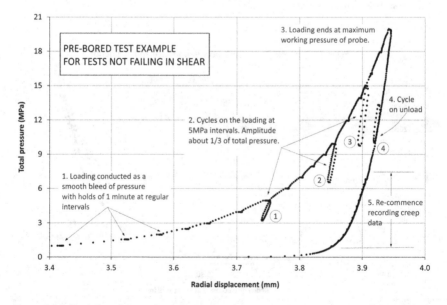

Figure B.19 Template for an elastic test in rock.

approximately 10 distinct steps in each of the unloading and re-loading parts.

5. Given three or more cycles of similar form, it is possible to determine the stress dependency of the modulus.
6. Creep readings continue to be taken throughout the loading but not during the cycles. Once above any plausible *in situ* stress level, the spacing can be increased, but not more than 1 MPa.
7. The test stops when the maximum working pressure of the equipment is reached, in this case 20 MPa. A final creep displacement is recorded then the cavity is unloaded, without gathering creep data.
8. Cycle 4 on the unloading is arranged to cover a similar stress range as cycle 3.
9. The cavity contraction continues without interruption until a point is reached that is above any of the principal *in situ* stress values (see note 3, above). The remaining stress is apportioned into 10 steps and creep readings recommence. This is a version of the balance pressure check, but carried out without any of the material having gone plastic. Unlike the initial creep readings, all fractures will now be closed and this makes it easier to identify points of minimum creep and possibly significant stress.

B.7 PUSHING AND REAMING

Inserting a pressuremeter into the ground by pushing into virgin soil is feasible but means that the device has to be relatively slender to minimise the forces required. The pressuremeter shown in two arrangements in Fig. B.20 is only slightly greater in diameter than a 15 cm^2 cone penetrometer (CPT). It may be placed using standard CPT trucks and can be arranged to carry a "live" CPT on the front end.

In principle, a combined device offers the speed and descriptive qualities of the cone with the parameter determining abilities of the pressuremeter test and such a device is called the cone pressuremeter (CPM). In practice, this may not be the optimum way of using both tools together. In a cohesive material, it is generally faster to make an initial probing with a 10 cm^2 CPT and then following its removal, ream out the hole with the pressuremeter, making tests at layers identified from the cone data.

When the additional parts for the live cone are removed, the probe is fitted with a dummy point and is much shorter. The material has already been pushed by the passage of the CPT and the pressuremeter is reaming out the existing hole to a larger diameter. Used in this way, it is referred to as a reaming pressuremeter (RPM).

The RPM can be used as a straightforward pre-bored pressuremeter, the hole being made by a small 50.8 mm core barrel or a small drag bit. However, the reduced length means that the probe can be used in an unusual way. If Standard Penetration Test (SPT) tools are available, then the probe can take

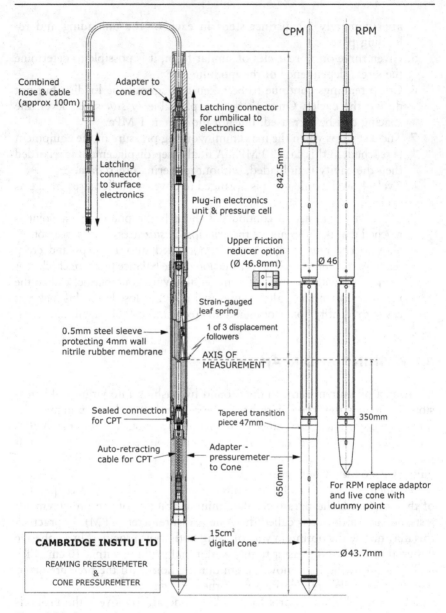

Figure B.20 Cone and reaming pressuremeter.

advantage of an SPT hole to carry out a cavity expansion. The material has been pushed by the SPT and the subsequent unloading puts the ground into a complex state, but the expansion range of the RPM is sufficient to erase most of the SPT-associated stress history.

Unload/reload cycles will be a valid sources of stiffness data and the final cavity contraction will give strength. Combined with a balance pressure check (BPC), the final contraction may also provide a reasonable estimate of the insitu lateral stress. Fig. B.21 is a template for this method of testing; Fig. B.22 is an expanded view of the BPC phase of the test.

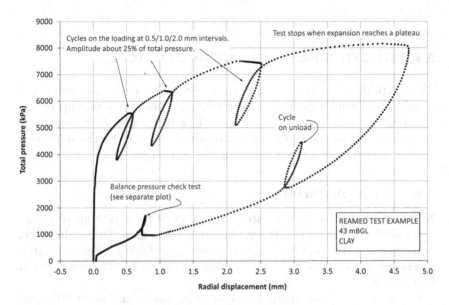

Figure B.21 Pattern for a test in an SPT pocket.

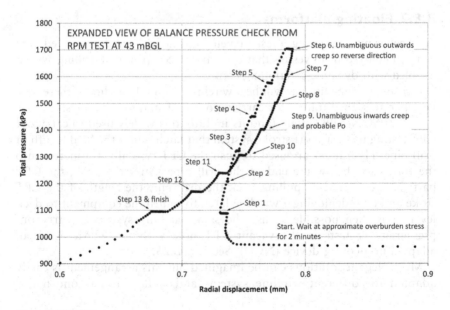

Figure B.22 BPC test.

By this means, a pressuremeter test can be obtained in a variety of situations and materials where it would be difficult to deploy the larger instruments. Because the loading phase of the test is largely inaccessible to analysis, it is never the preferred option but it may be the only feasible choice.

B.8 DEPLOYING PRESSUREMETERS OFFSHORE

The practical issues with using a pressuremeter offshore are concerned with speed of deployment and movement of the platform from which operations are conducted. As almost all testing will be referenced to the seabed or mudline, it is important to know the depth of water above the submerged surface (refer to the note on calculating k_o in Chapter 3).

B.8.1 Fixed platforms

Nearshore and shallow sea testing can be done from jack-up platforms using similar deployment systems to those used on land. This is almost the only system that makes it possible to carry out offshore self-bored testing unless the material is loose enough for jetting to be effective (refer to Chapter 5).

It is unusual for the platform height to be more than about 50 metres above the submerged surface.

B.8.2 Floating platforms

For deeper marine testing, pressuremeters of the pre-bored or pushed type can be deployed from vessels that use a heave compensated drilling system coupled with dynamic positioning control.

If casing is being used, it will be a wireline system. In addition, there may be separate seabed frames for carrying out CPT and sampling tests.

The WISON-APB system (and its imitators) is widely used to carry out CPT testing using a short stroke system that latches into the final length of drill pipe. Hydraulics are used to push the CPT typically 1.5 metres into the formation below the end of the drill pipe. After the push, the CPT parts are recovered by pulling back the drill pipe the length of the CPT stoke and then deploying a wireline latch to return the equipment to deck level. It is then possible to use the same latching system to carry out pressuremeter tests in the vacant CPT hole with the RPM, using an adapted non-coring device (NCD). See Fig. B.23.

More elegant solutions can be imagined but this arrangement is easily adapted for different wireline systems and is fast. It has one major

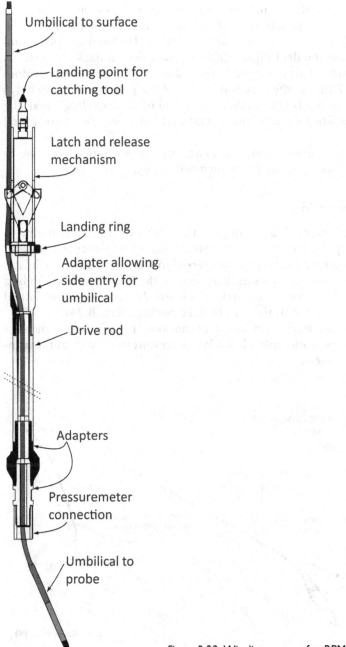

Umbilical to surface

Landing point for catching tool

Latch and release mechanism

Landing ring

Adapter allowing side entry for umbilical

Drive rod

Adapters

Pressuremeter connection

Umbilical to probe

Figure B.23 Wireline system for RPM.

disadvantage. In order to obtain the clearance for locking in the pressuremeter the drill pipe has to be lifted back at least one metre above the base of the borehole. The pressuremeter then latches into place and protrudes below the drill pipe, which is now lowered back down to its previous position. This is a delicate procedure. If the pressuremeter does not find the CPT hole, then the full weight of the pipe can act on what is a slender tube and it is easy to do a great deal of damage. In general, the system is unsatisfactory in granular material but works reasonably well in clays.

For operating in granular material, currently there is little alternative to deploying the pressuremeter on its own drill string.

B.8.3 Wave action

Ocean currents and swell are acting on any offshore surface attempting to keep stationary. The drill pipe or casing column will oscillate a small amount and some of this will be transferred to the pressuremeter whilst it is carrying out a test. It shows itself as noise in the field curve and can be a particular problem for unload/reload cycles. At relatively low lateral loads in weaker material, this can be very obvious (Fig. B.24).

At higher stress levels and larger expansions, the effect is somewhat mitigated. It is less noticeable when a larger pressuremeter such as the high-pressure dilatometer is used.

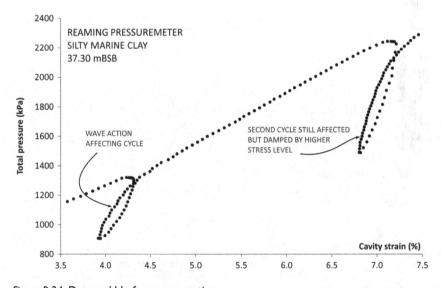

Figure B.24 Data wobble from wave action.

B.9 LOSSES

Even the simplest of the pressuremeters described in this appendix is a complex and relatively high-cost item. Preservation of the probe is an important consideration.

It may be thought that the self-boring probe is the most vulnerable to loss but the experiences of the manufacturer (Cambridge Insitu, CI) suggest otherwise. There have been a number of incidents where the downhole probe has become stuck, generally some distance below the cased part of the borehole. While a test is being carried out, loose material has an opportunity to settle on the top of the probe, making a ground anchor. Before this is fully appreciated an over-enthusiastic drilling crew has managed to break the outer 50.8 mm casing in an effort to pull the probe back.

Of the five times that this has occurred, the probe has been recovered but on two occasions the process damaged the instrument beyond economic repair. Four of these events happened in dense sand. The one case when the material was in clay was due to the cased borehole itself following a curved path and the sticking point was catching the top of the probe on the base of the casing. The probe was recovered successfully with purpose designed fishing tools.

Generally the SBP does not die but may become obsolete when the quantity of later innovations makes it impractical to attempt a retrospective upgrade. The longest-serving SBP currently being used by CI is affectionately known as "Dougal" and is 30 years old at the time of writing.

The reaming pressuremeter (RPM) is more likely to become bent than lost. The slenderness and length of the device makes it vulnerable when a large load is applied to the unsupported instrument. Fortunately, it is relatively simple to straighten a moderately bent RPM probe.

There is a lost probe in Holland, when the device was configured as a cone pressuremeter (CPM). In this case, the CPM was being pushed by a small track mounted crawler and had encountered a stiff layer of sand at some depth. The material near the surface was a very soft clay and one cone rod broke its thread some distance down as the string of rods deflected. Time constraints meant that no recovery could be attempted.

For most of the cases mentioned so far, if adequate casing had been used to support the borehole the problem would not have occurred.

The high-pressure dilatometer (HPD) has suffered the greatest number of losses, but it also does by far the greater number of tests. It is often used horizontally from within tunnels and it is then difficult to recover if it becomes stuck. Over-coring the probe tends to be the usual method and this inevitably results in significant damage. Three probes have been written off due to this cause.

There is a lost HPD off-shore Turkey when the dynamic positioning system broke down and the ship drifted from position, breaking off the drill pipe with an HPD installed in a pocket. There are also two prototype HPDs lost in the same 600-metre deep borehole in Israel. In this case there was a fault zone leaking material into the uncased borehole. For normal drilling this was not a difficulty because the near continuous passage of drilling tools prevented a buildup of material. Problems only arose when the operation was paused for long-duration pressuremeter testing. It took some time to appreciate the cause and the HPD was redesigned to mitigate the effects of this occurrence.

NOTE

1 "Kentledge" may be a term unfamiliar to some readers. It is the weight or resistance available to oppose a force being applied in the reverse direction.

Calibrating Cambridge Pressuremeters

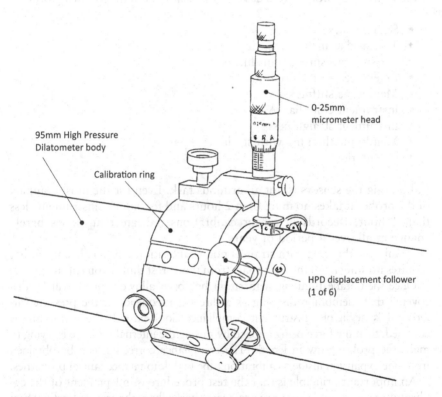

95mm High Pressure
Dilatometer body

Calibration ring

0-25mm
micrometer head

HPD displacement follower
(1 of 6)

Figure C.0 Calibrating an HPD displacement measuring system.

C.I INTRODUCTION

The purpose of this appendix is to demonstrate how the performance of the instruments is proved. The methods are specific to a particular kind of pressuremeter although the principles apply to all.

There are two kinds of calibrations. The most straightforward are those related to the performance of the transducers used to derive changes in stress and displacement. However, no matter how accurate the measurement, what is being determined are changes occurring at the *inner* surface of a sealed membrane. The purpose of the test is to determine the alterations to the cavity wall at the *outer* surface of the membrane. The second aspect to the pressuremeter calibration is calculating the difference between these two surfaces. The disparity is due to the strength of the membrane (significant in low strength ground) and the compliance of the instrument (significant in high stiffness material).

All Cambridge pressuremeters use similar procedures so examples are given from a variety of probes. This description is laid out in the following order:

- Scale factors
- The displacement measuring system
- Pressure measuring transducers
- Reference ("zero") outputs
- Membrane stiffness
- Instrument compliance
- Instrument straightness
- Membrane thinning with strain
- Orientation

Calibrating the sensors is not an arduous task. Even for the most complex of the probes it takes no more than 2 hours and for simpler instruments less than 1 hour. Records of sensor calibrations indicate that values barely change at all over a period of years.

Membrane thinning with strain is not a calibration, it is a calculation that requires nothing from the user apart from the at rest dimensions of the probe.

Once membrane thinning with strain has been allowed for, it will be discovered that membrane thinning as a direct consequence of the pressure differential is negligible. Instrument compliance does exist. If the membrane is confined, then the force being applied by raising the internal pressure is trying to make the probes grow in length. The compliance correction can be obtained from one expansion inside a calibration tube with known mechanical properties.

An important principle is that the test procedure is independent of the calibration procedures. There are pressuremeters where the test and calibration procedures mimic each other, and data are corrected on a point-by-point basis. This is an unnecessary restriction for the pressuremeters being described here.

C.2 SCALE FACTORS

The transducers in the probes are based on full bridge strain gauge circuits, where one-half of the bridge is in compression and the other is seeing

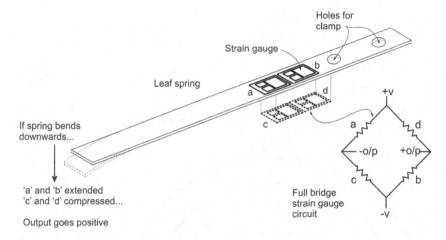

Figure C.1 A strain-gauged leaf spring.

tension (Fig. C.1). The gauges are matched to themselves and to the substrate on which they are mounted so that the effects of temperature alteration cancel out. The transducer produces an output dependent on the voltage being applied to it, the stress deflecting it and the amplification or buffering between it and the recording system.

The instruments contain electronic devices that provide a regulated voltage to the transducers and amplification of the resulting output signals. Because this electronic conditioning is a fixed part of the system, it is not mentioned when presenting calibrations. The electrical output of the transducer, in volts, is quoted only as a function of the deflecting stress. This function is termed "sensitivity" and gives the scale factor for deriving pressure or displacement from the transducer electrical output.

Although the output of the transducers is analogue, the final output of the system is a digital data stream of ASCII-encoded numbers representing volts. All variables associated with producing the digital output from the strain gauge signals are a function of the pressuremeter itself, and are independent of external changes (such as alterations to the connecting cable).

When using the sensitivity calibrations to convert readings from volts into engineering units, two important assumptions are made about the output; that it is linear and that the hysteresis is negligible. The calibration procedure is the evidence that these assumptions are reasonable.

C.3 THE DISPLACEMENT MEASURING SYSTEM

Because of their appearance the displacement measuring devices are often referred to as "the arms." The arms are calibrated by mounting a micrometer

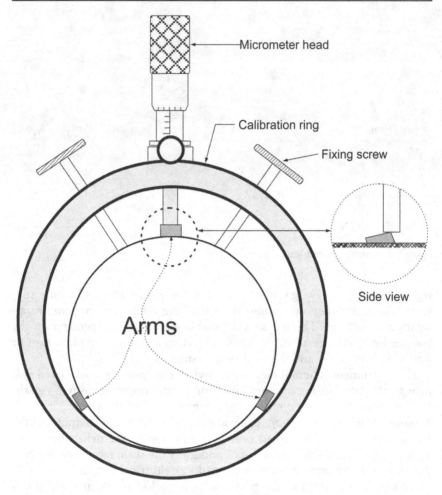

Figure C.2 Arm calibration jig.

above each in turn and recording the output for a given deflection (Fig. C.2 and Fig. C.0). When calibrating an instrument it is necessary to plot these readings for both an increasing and reducing deflection. The difference at a given point between the go and return path is a measure of the hysteresis. The worst case figure is noted, and corrective action is taken if the hysteresis is outside an acceptable limit, normally 0.5% of the sensitivity. The use of micrometer heads with a non-rotating stem gives the optimum hysteresis measurement.

The slope of the best-fit straight line through all the points is used to quote the arm sensitivity—typically as an output for a given deflection in units of millivolts per mm (Fig. C.3). It is not possible to see the hysteresis on such a plot and this is calculated from a spreadsheet table of the measured data (Fig. C.13).

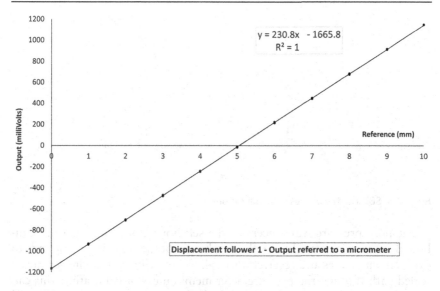

Figure C.3 Arm calibration example.

C.4 PRESSURE-MEASURING TRANSDUCERS

For pressure-measuring circuits, the maximum possible sensitivity is desirable; the only requirement is that the sensitivity be known and be linear and stable.

The sensitivity of *internal* pressure transducers is determined by placing a close-fitting large metal cylinder over the probe and applying a known pressure to the inside of the instrument (Fig. C.5). The pressure being applied is measured by a standard test gauge or similar device. As with the displacement followers, readings are plotted, the hysteresis noted and the best-fit straight line drawn through the plotted points.

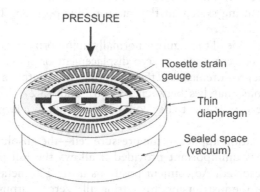

Figure C.4 Total pressure cell.

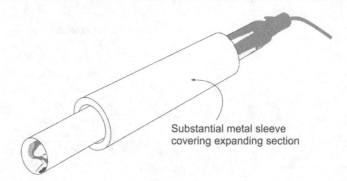

Figure C.5 SBP ready for pressure calibration.

The pore pressure transducers of the self-boring pressuremeter are calibrated in the same way. In use, the diaphragm of the transducer responds to external water pressure received through a filter. Fitting the filter creates a sealed path through the pressuremeter membrane. For calibrating, this can be awkward, because an externally applied pressurising system has to seal to the outer surface of the membrane. An easier alternative is to fit a temporary membrane without the passage to the filters so that the pore pressure transducers become, in effect, total pressure transducers and see the internal pressure.

Pressure sensitivities are quoted in units of millivolts per MPa and are arranged to be large enough to resolve a kPa of alteration to one decimal place.

C.5 REFERENCE ("ZERO") OUTPUTS

The other parameter that the transducers have is a known output for an "at rest" position. This is the value of the outputs produced by the circuits with atmospheric pressure both inside and outside the instrument, and any displacement measuring system at the initial radius position. This is called a little misleadingly "zero."

The absolute value of this figure is normally unimportant; it is not necessary that the figure be zero volts for zero displacement or stress, just that it be known. Current pressuremeters of the Cambridge kind output a digital number representing volts that lies between −3.2767 and +3.2767. The "at rest" readings for the arms are set to be about −2 volts to allow a large output range with a margin.

A similar situation applies to the pressure cell—the absolute value of the "zero" output is unimportant provided it allows the full pressure of the system to be resolved. Adjustment positions using 1% metal film resistors are provided in the instruments for setting all "zero" outputs.

It is recommended to take (and make a record of) zero readings both at ground level and also at test depth immediately prior to carrying out a test. A significant change between zero readings must be investigated. "Significant" would mean a change of 30 millivolts from the last set of zero readings. It is not unusual for shifts of a few millivolts to occur from day to day. It is important that the zero readings be stable when viewed over a period of a few minutes.

When using oil to inflate a probe, ground-level readings are the preferred reference because once in the borehole, the pressure transducers will read the head of oil. For gas inflation it is probably better to use the initial readings when the probe is in place in the borehole, because it will then be at the temperature most applicable to the test. However, the self-boring probe is fitted with pore pressure transducers. These two transducers must use ground-level readings as a reference if they are to read the full water pressure in the ground.

C.6 MEMBRANE STIFFNESS

The membrane that is expanded by a pressuremeter has its own initial tension, hence a small amount of pressure is required to expand it. The readings taken from the stress cells need to be reduced by this pressure to determine the net stress being applied to the ground.

The membrane correction is idealised as two components—a fixed pressure (in kPa) to move the membrane from its position at rest on the instrument, and a variable pressure dependent on the radial expansion, in units of kPa/mm (Fig. C.6). The technique for obtaining the correction data is to pressurise the instrument in free air.

The slope and the intercept on the pressure axis of the graph produced by this test give the membrane correction information for each arm.

The membrane does not necessarily possess isotropic properties and potentially a different set of calibration constants could be applied for each displacement follower. Fig. C.7 is the same test as Fig. C.6, but with the trend of individual displacement followers plotted. It seems wildly anisotropic but what this indicates is a persistent problem when calibrating, which is creating a reasonable representation of conditions in the ground. An unconfined inflation in air exaggerates any variation in membrane properties because it is unresisted. The weakest part of the membrane dominates the expansion and the strongest part the contraction to an extent that is not possible if the surrounding medium has strength and stiffness. The consequence is that in air the membrane response looks like a hysteresis loop. The nearest approximation to the behaviour of the membrane in the ground will be the unloading phase of the test in air.

An additional problem is that without the ground to contain the expanding membrane, its axis moves in relation to the axis of the measurement system.

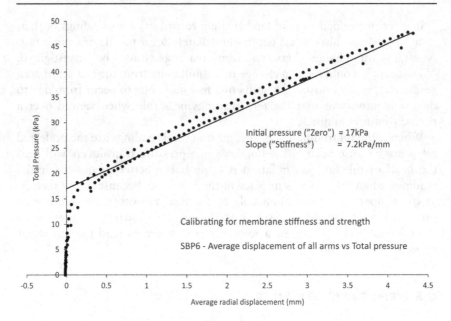

Figure C.6 Membrane calibration, average displacement.

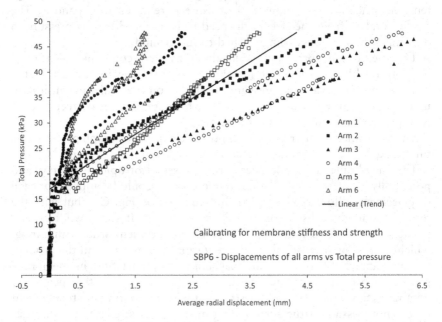

Figure C.7 Membrane calibration, individual axes.

The instrument used to make Fig. C.6 has six regularly spaced displacement followers. This makes it possible to look at the average radial displacement across three diameters, which largely cancels out this effect. This is done in Fig. C.8 and gives an identical result to the average of all arms. The

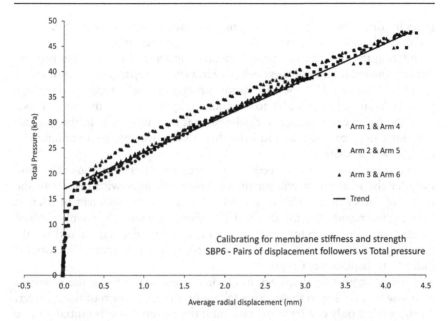

Figure C.8 Membrane calibration, diametrical displacements.

conclusion is that only the results of the average correction from this procedure should be used to correct the raw pressure data.

Correction figures for rubber membranes about 4 mm thick might be 30 kPa and 15 kPa/mm. The SPB membrane in the example is only about 1 mm thick and made of polyurethane. This gives the lowest slope correction.

C.7 INSTRUMENT COMPLIANCE

The instruments will deform as a consequence of the large pressure being internally applied. Put simply, the probes stretch. The displacement measuring system uses the body of the instrument as a reference so axial movements of the body are seen as small but measurable displacements of the membrane. This system compliance has implications for the determination of shear modulus, and will become a significant source of error when measuring shear modulus values in the gigapascal range.

There are a number of effects to consider but they are collectively determined using a single procedure. The correction figure which results is known for historical reasons as "membrane compression."

This is not a good description.

For the Cambridge family of pressuremeters, real membrane compression, that is the membrane changing in thickness as a direct result of the pressure differential across it, is almost too small to be measurable. There are a number of other factors to consider of greater impact.

Inflating the instrument inside a metal cylinder will in principle provide data on the magnitude of these effects. However, a separate source of error, which is a function of the calibration procedure itself, then becomes apparent. The membrane is able to expand axially by a small amount and, as a result, experiences a change in thickness that may not occur in the ground. Although steps can be taken to keep this axial movement to a minimum, it cannot be eliminated.

As a consequence of the freedom of movement offered by the calibration cylinder the instrument will move, its centre will alter with respect to the centre of the cylinder. This is not helped by the low coefficient of friction between the membrane and the steel by comparison with the membrane and the ground. Only average radial movement can be derived from this calibration process, and it is not sensible to try to obtain separate factors for individual displacement followers.

If metal protective strips are fitted then the initial part of the observed response will be due to these strips taking up the exact form of the cylinder, a process that only occurs in the ground if the material is substantially more stiff than the instrument. This is the explanation for much of the initial curvature that occurs when an assembled probe is inflated inside a metal sleeve; it is a serious error to attempt to derive a correction factor from this part of the loading.

Taking account of all the above, the following method is used. The outer sheath is removed and a thick wall metal tube of known properties is placed over the membrane of the instrument. The instrument is pressurised until the membrane contacts the wall of the tube. The test continues, either as a gentle continuous inflation or in discrete steps, holding each step briefly to ensure maximum accuracy. The probe is taken to maximum working pressure. The pressure is then reduced using the same pressure steps. Ten such steps covering the full working range are sufficient. Some users prefer the unloading should stop short of zero, and then the probe should be reloaded again to maximum pressure and unloaded to zero, in effect doing a large unload/reload cycle. In a good calibration, all loading and unloading slopes will be similar, but if the probe moves with respect to the cylinder, this will affect the data. In this event a second reloading gives the best correction information.

The calibration is obtained by plotting the pressure/displacement data on an expanded scale, and finding the best-fit slope through the points. The slope ideally should be the calculable expansion of the cylinder for the pressure applied. In practice it is always more, the difference being the compliance. The units are "mm/GPa," a typical value being between 3 and 6 mm/GPa, depending on the type of probe. See Fig. C.9.

The properties and dimensions of the calibration tube are a part of this calculation and its contribution will need to be calculated using thick wall cylinder theory (Lamés equations).

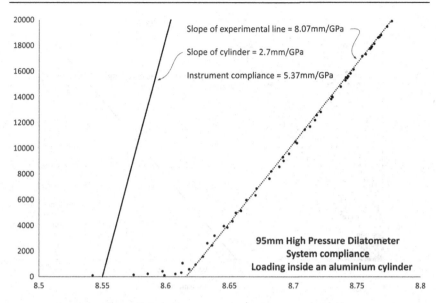

Figure C.9 System compliance calibration.

Quoting the compliance in this manner allows the software to calculate the appropriate error for every step of pressure and to make the necessary adjustment to the measured displacements.

To put the correction in context, the correction in the example is equivalent to a shear modulus of 4.5 GPa. The calibration is highly repeatable, within about 5%, so shear modulus of 20 GPa or more can be determined within acceptable limits.

A particularly severe examination of the displacement system can be obtained by removing the lever part of the arm follower system so that the strain-gauged leaf springs have no mechanical role to play. If the "compliance" calibration is now carried out the result ought to be a vertical line if the transducer were perfect. Fig. C.10 is an example of what is obtained in practice. The apparent movement is any remaining compliance in the measurement system added to the imperfect nature of the transducer—to some extent the transducer is pressure sensitive. As is evident with comparison to a shear modulus of 10 GPa, the errors are small. Provided that no changes are made, the errors will be a constant characteristic of the displacement measurement system.

C.8 INSTRUMENT STRAIGHTNESS

The pressuremeter can become bent during operations due to the large forces that occasionally are applied before the probe is supported by the borehole. The straightness may need to be checked.

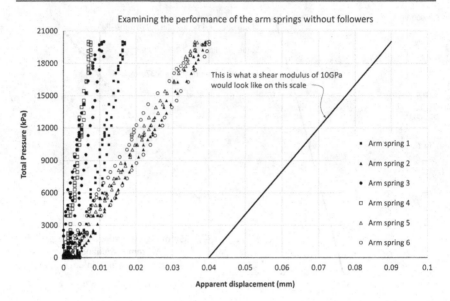

Figure C.10 Pressure sensitivity of displacement system isolated.

The method for doing this is to support the instrument at the points where the membrane is clamped, and then to rotate the instrument whilst the run out is observed at a number of points along its length. A record is kept indicating the total runout at these points, and the probe can be straightened if the eccentricity is significant. An acceptable discrepancy over the measuring section is 200 μm.

The instrument is never perfect, and it happens that frequently a consistent bias in the displacement system (especially in the vicinity of initial movement of the membrane) can be linked to a lack of straightness.

This is an important check for the self-boring probe.

C.9 MEMBRANE THINNING WITH STRAIN

During a test, the pressuremeter membrane changes in thickness as a consequence of being stretched. This change in thickness can be calculated, assuming that the cross-sectional area of the membrane remains constant. The calculation is incorporated into the program that converts raw data into engineering units.

The term "membrane" includes the stainless-steel protective sheath. The measurement that initiates the calculation is the radial distance to the inside surface of the membrane (*d* in the definitions given below).

Definition of Terms

a	is the internal radius of the membrane at rest
b	is the external radius of the membrane at rest
c	is the internal radius of the membrane expanded
r	is the external radius of the membrane expanded
t	is the thickness of the stainless-steel sheath strips
d	is the measured movement of the strain arm
E	is the actual expansion of the outer surface of the membrane

Calculation

At rest, the cross-section area of rubber $= \pi(b - t)^2 - \pi a^2$

The expanded cross-section area of rubber $= \pi(r - t)^2 - \pi c^2$

Because the membrane is incompressible unless confined, these must be equal:

Therefore $\quad (b - t)^2 - a^2 = (r - t)^2 = c^2$

Now $\quad c = a + d \quad \& \quad r = b + E$

Hence $\quad (b - t)^2 - a^2 = [(b + E) - t]^2 - (a + d)^2$

$$E = \sqrt{[(b - t)^2 + d(2a + d)]} - (b - t) \qquad \text{[c.1]}$$

$2a$ and t are known from the manufacturer's data. $2b$ is also given data, but can be confirmed by measuring the external diameter of the instrument concerned.

Typical dimensions for three types of probe are:

	47 mm RPM	88 mm SBP	95 mm HPD
	(mm)	(mm)	(mm)
$2a$	38.0	79.2	81.0
$2b$	47.0	88.1	95.0
t	0.5	0.5	0.5

To apply the correction at a given expansion, the *average* radius of the expanding membrane is calculated. This average is then entered into the equation and the ratio between the corrected average and the raw average is expressed as a scale factor (for example it is about 0.86 for an RPM at all expansions). The scale factor is then applied to the individual arm displacement outputs.

C.10 ORIENTATION

The high-pressure dilatometer carries an electronic compass at the lower end, able to determine the orientation of the probe. The compass consists of

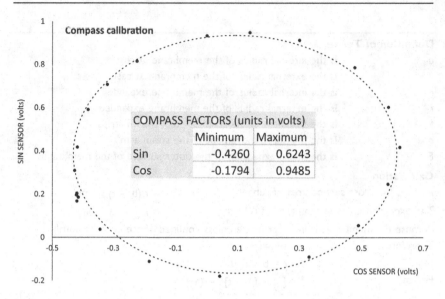

COMPASS FACTORS (units in volts)		
	Minimum	Maximum
Sin	-0.4260	0.6243
Cos	-0.1794	0.9485

Figure C.11 Compass calibration.

two flux gate magnetometers fitted at right angles to each other. These give the sine and cosine of the angle made with magnetic north, permitting an unambiguous direction to be derived.

To calibrate the sensors, the instrument is rotated slowly through 360 degrees whilst the output of the sensors are logged (Fig. C.11). From this, the maximum and minimum output of each sensor is discovered and is stored. Thereafter, the ratio of the two sensors gives the tangent of the angle, easily converted to a bearing.

A mark on the outside of the compass module indicates the position of the Cos sensor. The module can be arranged to fit to the instrument so that this stud is in line with axis 1. The direction that the compass produces is the bearing of arm 1 with respect to magnetic north.

In practice it is generally more straightforward to note the *mis*-alignment or offset of axis 1 with respect to the Cos sensor and introduce an adjustment later. It is also necessary to use third-party data to identify the declination at the current location so that the final orientation can be expressed as a bearing with respect to true north.

The calibration has to be carried out away from any metal such as drill rigs or casings.

C.11 APPROPRIATE APPLICATION OF CALIBRATIONS

At several places in this appendix, reference has been made to the difficulty replicating on the test bench what will take place in the ground. It is all too easy to apply poor calibrations to good experimental data.

Although it is important to regularly check the sensitivities of the strain gauge circuits, it is unusual for them to change markedly. Indeed, it is common for the hysteresis to improve with use or exercise. 90% of the performance of a strain gauge bridge application can be predicted from its design; the calibration removes the uncertainty due to manufacturing tolerances and can give early warning of impending problems in a particular circuit.

The expansion test is concerned with making relative, not absolute, measurements. The displacement measuring system will resolve movements of less than 0.5 µm over a range of 13 millimetres; the pressure measuring system will resolve changes of 0.1 kPa over a range of 20 MPa.

This resolution is considerably higher than can be seen with a standard micrometre or test gauge. 0.5 µm is approximately the wavelength of ultraviolet light.

Clearly there is no practical possibility of checking by measurement a movement so small. What is done in practice is to check that the various sensors are linear over a number of relatively coarse steps or intervals. It is assumed that this linear behaviour will be true for very much smaller changes. If an attempt is made to look at smaller changes with a measuring instrument, unexpected results can be obtained.

Fig. C.3 is a calibration for displacement follower sensitivity where a micrometer was used to apply steps of 1 mm. A change of 1 mm is two full rotations of the micrometer thimble. It is possible to look at smaller steps, such as 0.1 mm. The same displacement follower was calibrated in this way over a 1 mm range, increasing and decreasing. See Fig. C.12.

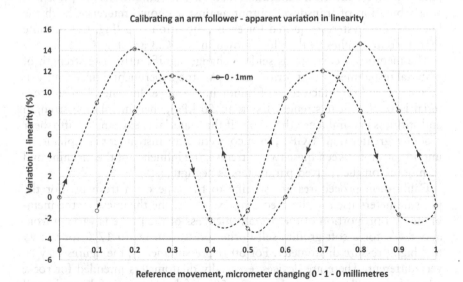

Figure C.12 Micrometer thread wobble.

The linearity of the earlier calibration appears to be lost, and what is now seen is something that resembles a thread form—which is precisely what it is. A micrometer is merely a nut running on a very fine thread and is only guaranteed to meet its specification after a full turn. The displacement follower is seeing the thread wobble of the measuring tool. In this instance, its resolution and linearity are considerably better than the device to which its output is referenced.

Smaller sources of error in this assumption of linearity, such as temperature change, are largely ignored. When critical measurements are being made during a test, for example when taking a reload loop, it is reasonable to assume the temperature remains constant. The ground is usually at a constant temperature whenever a test is carried out, but sometimes there are problems: the temperature of the gas being supplied to the downhole tool can have an influence especially if the gas bottle reservoir is lying outside in direct sunlight, a particular difficulty in locations such as Saudi Arabia. Under these circumstances, if the gas supply cannot be kept in a temperature-controlled environment, the instruments should be inflated hydraulically using a light oil.

Conversely, operating in frozen ground may mean that the assumption of a constant ground temperature is unsafe. Temperature then has to be recorded as part of the data stream and potentially used to modify calibration constants unless the measurements can be demonstrated to be immune to temperature effects over the range in question.

Spreadsheet software is used to present the results of the calibrations for sensitivity. One benefit of this is that gradients are calculated by linear regression routines; this ensures that different operators given the same set of data will derive identical calibration factors. The calibrations are presented as a tabulation of transducer output against a known reference, with the linearity and hysteresis quoted for each calibration step (Fig. C.13). Note that the data applies to the plots shown in Fig. C.3 and Fig. C.12.

The membrane corrections seldom change significantly. The strength of material tested means that for the most part the experimental uncertainty in deriving membrane factors is of minor importance. In general, if the material is weak (shear strength less than 50 kPa), then membrane strength and stiffness can influence the result. If the material is extremely stiff (shear modulus greater than 1 GPa), then correcting for instrument compliance is important. In between these two extremes, the influence of the machine and membrane on the derived parameters is negligible.

Calibration procedures are specific to the type of instrument. For the pressuremeters used in the preceding examples, the thickness of the membrane as a proportion of the overall thickness of the probe is varying from about 4% for a soft ground self-boring pressuremeter to 17% for the 95 mm high-pressure dilatometer. For other pressuremeters, the figures may be very different. There are devices, especially dilatometers intended for rock, where the rubber membrane represents 50% of the diameter of the probe and this difference will heavily influence calibration procedures and testing.

0-10mm	Arm (mVolts)	Linearity (%)	Hysteresis (%)	0-1mm	Arm (mVolts)	Linearity (%)	Hysteresis (%)
0	-1162.0	100	0.17	0	-1162.8	94	-0.39
1	-930.8	100	0.28	0.1	-1140.9	103	-0.26
2	-700.7	100	0.31	0.2	-1116.9	108	-0.17
3	-469.5	100	0.35	0.3	-1091.7	103	0.09
4	-239.6	100	0.34	0.4	-1067.6	92	0.22
5	-9.3	100	0.35	0.5	-1046.2	93	0.30
6	222.1	99	0.39	0.6	-1024.6	103	0.17
7	451.7	101	0.33	0.7	-1000.5	106	0.34
8	684.0	100	0.39	0.8	-975.8	102	0.09
9	915.9	100	0.23	0.9	-952.0	92	0.09
10	1147.4	103		1	-930.5	93	
9	910.7	102		0.9	-952.2	102	
8	675.0	100		0.8	-976.0	109	
7	444.1	100		0.7	-1001.3	102	
6	213.2	100		0.6	-1025.0	94	
5	-17.3	100		0.5	-1046.9	91	
4	-247.5	100		0.4	-1068.1	102	
3	-477.5	100		0.3	-1091.9	106	
2	-707.9	99		0.2	-1116.5	102	
1	-937.2	99		0.1	-1140.3	93	
0	-1165.9			0	-1161.9		
Zero	-1165.8 mV			Zero	-1162.8 mV		
Slope	230.8 mV/mm			Slope	232.9 mV/mm		

Figure C.13 Table of displacement calibration readings.

If the pressuremeter is a volume-measuring device, then the adjustment for membrane thinning no longer applies—the volume of ground displaced will be the same as the change of volume inside the probe. This is a theoretical advantage only, because such systems tend to require calibration for pressure-dependent volume changes in the umbilical. For the instruments described here, where measurement and signal conversion are implemented in the probe itself, changes to elements external to the probe, such as the connecting hose and cable, are irrelevant.

Appendix D

Comparing the Strain Distribution Around a Cone Penetrometer and Pressuremeter

Figure D.0 The cone experiment calibration chamber.

D.1 INTRODUCTION

There is a tradition in engineering of simple domestic-style experiments using easily available items to illustrate complex problems. This is a contribution.

The common *in situ* testing tools for site investigation in Europe, the United Kingdom and North America are the electric piezometric cone (CPT), initially developed in Holland, and the pressuremeter (PMT), initially developed in France. There has been extensive research applied to

303

both, in the hope that the data they produce can be resolved into engineering parameters for strength and stiffness.

Of the two, the cone with its rapid and relatively easy deployment has seen the greatest use. It is an excellent tool for providing a detailed lithology of a site and can delineate accurately layers of sand, silt and clay. It is also reasonably straightforward to use the cone as a model for a pushed pile and relate cone tip load to ultimate bearing capacity. With the addition of a seismic sensor the vertical seismic velocity can be determined and estimates for the elastic shear modulus obtained. However, despite extensive efforts, it has not been possible to turn conventional CPT output into values for modulus or strength except by empirical methods.

The pressuremeter is a slower and often more complex procedure to apply and does not reflect the detailed lithology[1], but provides fundamental engineering properties appropriate to the design of large structures. These are obtained by direct analysis of the measured field curve. The analyses are dependent on the stress path followed (drained or undrained) but in all other respects are independent of the material and location. Unusually for site investigation procedures the derived parameters can be used to reproduce the experimental data. A parameter set is a coherent representation of the material behaviour for a particular stress path and there is minimal requirement for additional input from third-party data.

This note considers some of the reasons for this basic difference by creating pictures of the strain/displacement of sand surrounding both a cone and a pressuremeter. In the past, x-ray techniques have been used to do this. The technique used here is to consider a "half" cone sliding on a glass surface. The half cone is surrounded by pseudo-sand (quinoa seed). Photographs of the cone as it is pushed through the material show the complex movement of individual grains. This is then compared to the simpler movement of an expanding pressuremeter. Despite the simplicity of the experimental setup, it is sufficient to show why one type of test lends itself to mathematical analysis, and the other does not.

Fig. D.1 shows that granular material is the biggest challenge for the CPT because this has the effect of eliminating two out of the three potential data sources for the standard cone. There is little or no response from the friction readings and the piezometer shows only the standing head of water—there is no excess pore pressure. Hence, in such material, the CPT outputs one thing only to represent the material behaviour, namely the tip stress.

D.2 EXPERIMENTAL SETUP

The first experiments carried out used a CPT cut from wood. The experiments have been repeated using a "half" cone machined from steel with similar results. To make the "sand" grains easier to identify, white quinoa

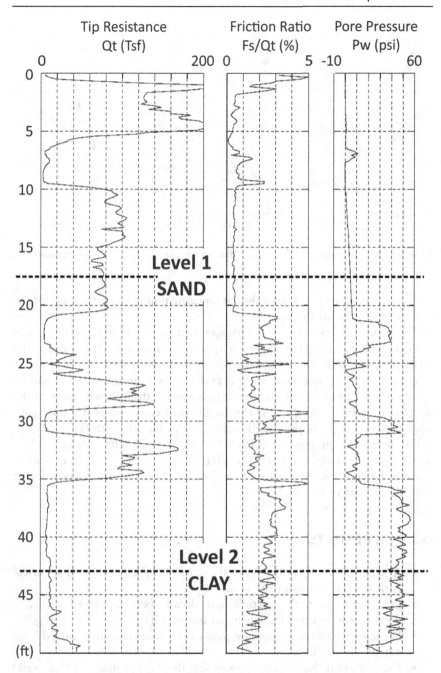

Figure D.1 Cone profile in sand and clay.

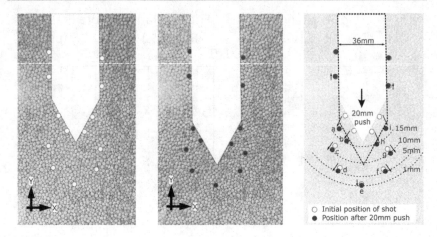

Figure D.2 Initial arrangement.   *Figure D.3* After 20 mm push.   *Figure D.4* Particle movement.

seeds are used, laced with lead shot at strategic points along the sides of the cone and in front of the tip.

The cone and lead shot are photographed in the at-rest position, then after an advance of 20 mm. By comparing before and after images, the movement of the shot can be seen, measured and converted to strain.

These results are not to be treated as definitive in magnitude but will be representative of direction and very approximate strain levels. Despite the simplicity of the arrangement, the results and conclusions are also consistent with previously published work (Teh & Houlsby, 1991).

For reproduction purposes, Fig. D.2 and Fig. D.3 have been slightly manipulated to emphasise the position of the lead shot. Fig. D.4 summarises the changes observed when advancing the cone from its position in Fig. D.2 to that in Fig. D.3.

D.3 THE RESULTS

- A particle initially at the tip of the cone is pushed on a semi-spherical path the full radius of the cone.
- Previous to that, the particle had already been displaced vertically downwards as the cone advanced towards it.
- Particles not in the immediate path of the cone have also moved both axially and radially.
- Particles that have been passed by the CPT show a small axial movement upwards as the passage of the cone continues.

Ignoring the lead shot adjacent to the parallel part of the cone, the remaining nine pellets in Fig. D.4 have been labelled a–i. An arbitrary origin

Table D.1 Strain readings from model cone push

Position	Radial strain (X) (%)	Axial strain (Y) (%)	XY strain (%)
a	26.5	62.5	67.8
b	31.9	50.5	59.7
c	12.6	24.5	27.6
d	9.2	20.0	22.0
e	3.3	10.8	11.3
f	2.7	15.5	15.8
g	6.7	26.0	26.8
h	16.3	59.2	61.4
i	20.5	57.0	60.6

has been identified that remains unchanged between "before" and "after" images. The horizontal (X) and vertical (Y) movements of these nine particles relative to the origin are summarised in Table D.1. They have been converted to strain using the radius of the cone as the reference dimension.

An inspection of the data suggests:

a. The strains are very large.
b. Axial strains are greater than radial strains, so the deformations are neither cylindrical nor spherical.
c. Every particle is following a different path.

The only source of deformation information is knowing that particles initially aligned with the CPT axis are displaced to the radius of the cone as it advances. Because no soil is removed, this is total disturbance, essentially infinite strain.

This is useful knowledge; if it is possible to say that the Qt values represent the limiting condition for the material, then *potentially* the cone can provide a limit pressure that is the ultimate output of a cavity expansion solution such as Gibson and Anderson (1961). What it cannot do is give any information for the constituent parts that make up that solution. It is those parameters (strength, stiffness and initial stress) that give the solution the predictive capability for determining the state of stress and strain at all intermediate conditions below the limit state.

We know where we are but have no idea how we got there.

D.4 THE PRESSUREMETER

For the CPT, the insertion and the loading phase of the ground are the same thing. For the pressuremeter, they are separate events.

D.4.1 Installation

The simple experimental arrangement used for the CPT cannot reproduce the effects of a self-bored installation. X-ray studies carried out by Hughes (1973) show that a thin wall tube can be pushed into soil without significantly displacing the material a very short distance from the surface of the tube.

Sensors cannot be installed in a tube of the necessary thinness, so self-boring is a means of inserting a thick wall tube with thin wall consequences. The self-boring probe removes from the ground a volume of material identical to itself.

The usual alternative to self-boring is pre-boring. Wroth (1978) gives examples of a radiographic experiment where a thin wall tube was pushed into a block of kaolin, and then after 30 minutes was removed. Fig. D.5 is a

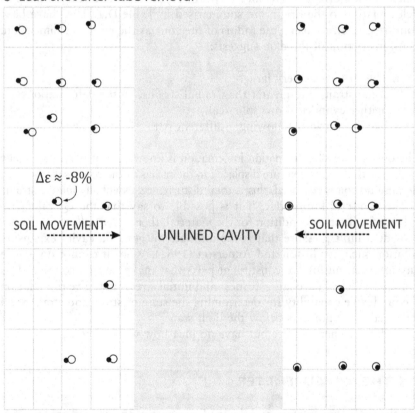

Figure D.5 Pre-bored strains (from Wroth, 1978).

composite of his images redrawn for clarity. The inward movement towards the unsupported cavity is substantial and illustrates why pre-bored testing in soft clay is not an attractive option.

D.4.2 Cylindrical cavity expansion

Using a similar setup to the CPT example (Fig. D.2), it is possible to mimic the cavity expansion test. The "pressuremeter" is a rectangular length of wood that takes up the length of the left-hand side of a mass of quinoa seeds. The pressuremeter is then pushed into the mass of grains a known amount, equivalent to an increase in radius of 25% (Fig. D.6).

Four data points have been selected. "R" is on the edge of the pressuremeter where it contacts the material. A, B and C are identifiable quinoa seeds at different radii within the "soil." The top image is the initial setup, and the lower image is the result following the simulated expansion. The point that was "A" has moved to "A*" and so on. The results are given in Table D.2.

The strain values are change of radius over current radius. The final column is the ratio of the radius at the particle to the initial radius of the pressuremeter.

Consider what happens at point "R." The before and after images give the measured displacement of the "cavity" inner surface. With this single

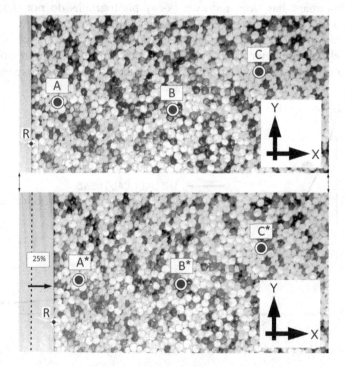

Figure D.6 Simulating a drained cavity expansion.

Table D.2 **PM results**

POINT	Δr/r (%)	r/r$_o$
R	25.0	1.00
A	18.4	1.29
B	3.7	2.55
C	0.8	3.49

measurement, it is possible to predict the movement at every other point in the soil mass. Fig. D.7 plots the data given in Table D.2. The theoretical trend is easily calculated because strain varies inversely with the square of the radius. This assumes constant area deformation, which will give inaccurate results in a drained material but is sufficiently close for the present purpose.

Given the simplicity of the setup and the number of theoretical considerations that have been violated, the agreement between experiment and theory is surprisingly good.

This predictability is due to all movement being in the radial direction, a finding that is consistent with the investigations reported in Hughes (1973). One of his x-ray images shows a pressuremeter expansion in a purpose made box where the material is normally consolidated dry sand[2] laced with lead shot. This image has been redrawn (x-ray photographs do not reproduce well) and is shown in Fig. D.8. The purpose of the experiment was to test whether the movement in the soil is indeed radial. Because of size constraints, the model pressuremeter has a length to diameter ratio of 2, about a third of the ratio of commercially available pressuremeters. The radial expansion is

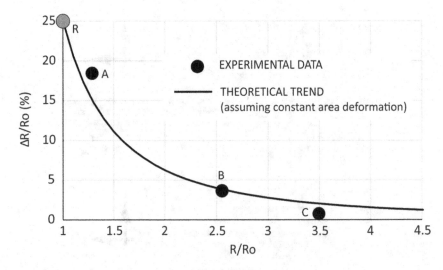

Figure D.7 25% cavity expansion: predicted and measured trend.

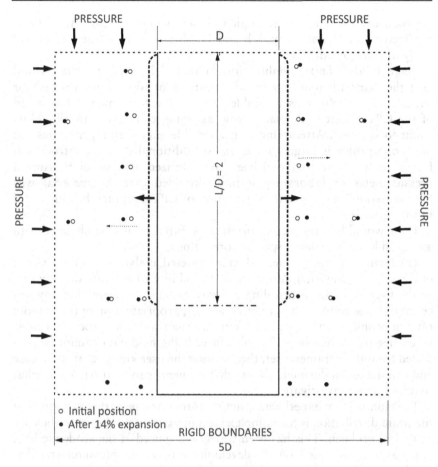

Figure D.8 Pressuremeter expansion (redrawn Fig 4.3 from Hughes, 1973).

about 14%. At the extremities of the expanding section, there is slight axial as well as radial movement but over the central third (where the displacement followers are placed), the movement is exclusively radial.

Even with this extreme geometry, the test is best modelled as a cylindrical cavity expansion. It also suggests that theoretical work with FEM that imply significant finite length effects should be viewed with some scepticism.

As Fig. D.7 shows, with a pressuremeter test we know where we are, we know how we got there and where we are going.

D.5 CONCLUSIONS

The resistance to penetration in a granular material is a function of at least four variables: the friction angle, density of the sand, the friction between the cone and the sand and the rigidity index (strain to yield). All of these parameters are identified by Teh and Houlsby (1991) in their analytical study of cone penetration.

There is an infinite combination of material properties that would give the same tip load, so for the purpose of identifying strength or stiffness, the problem is insoluble. This is the fundamental limitation of the CPT and it is the same issue as trying to convert an SPT blow count to stiffness. Attempting to get sensible engineering properties out of a cone profile is largely futile *unless* additional data are introduced; for example, a site that had been characterized at key points using a pressuremeter or laboratory testing. Provided that the material was homogeneous, an informed examination of CPT data may be able to fill in the gaps.

These would be *site*-specific correlations. What cannot be obtained with any confidence is *material*-specific correlations.

The feature of pressuremeter data interpreted analytically is the absence of a need for additional information. The ability to measure directly two of the four variables controlling a cavity expansion means that any uncertainty is a consequence only of an inappropriate assumption—more often a problem with tests in rock rather than soil. The proof of a pressuremeter parameter set is its coherence. If the field data cannot be predicted from the parameter set, then at least the user knows that is the case and can make a judgement about what elements can be trusted and what require further investigation.

The quinoa grain experiments, simple as they are, nevertheless show that the strain distribution is too complex for a fundamental solution to the CPT analytical problem. This has been widely recognized at the academic level and was the motivation for the development of a cone pressuremeter that combined features of both techniques.

A CPT is a close analogue for certain types of pile installation. Used in this way, the tip load bypasses fundamental considerations and the only remaining question is one of scale.

The same result can be obtained (more slowly) using a pressuremeter test. However, if the intention is to understand and model the interaction between piles, then a pressuremeter test is the only option.

NOTES

1 Although the shape of the loading curve gives an unmistakeable indication of the material being tested.
2 The miniature self-boring probe was jetted into the sand using compressed air as the flushing medium. It was some considerable time later before this method was attempted in field tests on a full sized probe using pressurised water in place of air.

REFERENCES

Gibson, R.E. & Anderson, W.F. (1961) In situ measurement of soil properties with the pressuremeter. *Civil Engineering and Public Works Review*, 56, 658, pp. 615–618.

Hughes, J.M.O. (1973) An instrument for the in-situ measurement of the properties of soft clays. Ph.D. thesis, Cambridge University.

Teh, C.I. & Houlsby, G.T. (1991) An analytical study of the cone penetration test in clay. *Géotechnique* **41** (1), pp. 17–34.

Wroth (1978). Cambridge in-situ probe. *Proc. Symp. on Site Exploration in Soft Ground Using in situ Techniques*. May 1–3, Alexandria, Virginia, pp. 97–135. Report No. FHWA-TS-80-202.

Index

Printed in the United States
by Baker & Taylor Publisher Services

Printed in the United States
by Baker & Taylor Publisher Services